Artificial Intelligence in Unreal Engine 5

Unleash the power of AI for next-gen game development with UE5 by using Blueprints and C++

Marco Secchi

Artificial Intelligence in Unreal Engine 5

Group Product Manager: Rohit Rajkumar
Publishing Product Manager: Bhavya Rao
Book Project Manager: Arul Viveaun S
Senior Editor: Mark D'Souza
Technical Editor: K Bimala Singha
Copy Editor: Safis Editing
Proofreader: Mark D'Souza
Indexer: Tejal Soni
Production Designer: Ponraj Dhandapani
DevRel Marketing Coordinator: Nivedita Pandey

First published: October 2024

Production reference: 1151024

Published by Packt Publishing Ltd.
Grosvenor House
11 St Paul's Square
Birmingham
B3 1RB, UK.

ISBN 978-1-83620-585-2

www.packtpub.com

To my nephews, Camila Mae Lynn Marcellina and Livio Siu Long, for, well... just being here.

To my students; you are the reason I keep on teaching.

And finally, to my adorable wife, Ambra – or should I say, "Professor Viktoria" – for helping me out with all the crazy ideas I come up with every day.

– Marco Secchi

Contributors

About the author

Marco Secchi is a freelance game developer who graduated in computer engineering at the Polytechnic University of Milan. He is a lecturer and lead game advisor at **Nuova Accademia di Belle Arti** (**NABA**) where he also mentors BA students in their final thesis projects. In his spare time, he reads a lot, plays video games (less than he would like), and tries to practice CrossFit.

About the reviewers

Syed Zaib Farooq is an accomplished entrepreneur and game developer with extensive experience specializing in gaming, VR, and immersive simulations. His expertise spans a range of industries, including real estate visualization, multiplayer gaming, metaverses, and EdTech simulations. Renowned for his proficiency in Unreal Engine, Zaib Farooq excels in developing custom plugins and utilizing both C++ and Blueprints. Driven by a passion for innovation, he consistently strives to create scalable gaming and simulation experiences that redefine immersion.

Nicholaus Price is a freelance game developer who has programmed in C++ since 2015 and in the Unreal Engine since UE3. His favorite genres to program are action, RPGs, platformers, and third-person shooters. Nick is currently an AI programmer for Project Sentinel. *Artificial Intelligence in Unreal Engine 5* is the first book that Nick has reviewed.

Table of Contents

Part 1: Introducing Artificial Intelligence in Games

1

Getting Started with AI Game Development   **3**

2

Introducing the Unreal Engine AI System   **15**

Part 2: Understanding the Navigation System

3

Presenting the Unreal Engine Navigation System   31

4

Setting Up a Navigation Mesh   47

5

Improving Agent Navigation   73

6

Optimizing the Navigation System   91

Part 3: Working with Decision Making

7

Introducing Behavior Trees   111

8

Setting Up a Behavior Tree   123

9

Extending Behavior Trees   159

10

Improving Agents with the Perception System 199

11

Understanding the Environment Query System 221

Part 4: Exploring Advanced Topics

12

Using Hierarchical State Machines with State Trees   235

13

Implementing Data-Oriented Calculations with Mass   273

14

Implementing Interactable Elements with Smart Objects   295

Appendix – Understanding C++ in Unreal Engine   313

Index   325

Other Books You May Enjoy   334

Preface

Since its very beginnings, **artificial intelligence (AI)** has transformed the landscape of game development, offering players enriched and immersive experiences that were once thought to be the realm of science fiction. Unlike other methods of game development, where scripted events dictate more rigid player interactions, AI introduces a level of unpredictability and responsiveness that brings virtual worlds to life.

In recent years, advancements in AI technology have made it more accessible and easier to implement, resulting in a surge of intelligent gameplay across various genres. This evolution has made AI a cornerstone of modern game development, with millions of players benefiting from smarter, more engaging interactions in their favorite games every day.

Unreal Engine stands out as an advanced platform for developers looking to harness the power of AI in their projects. With a robust suite of tools and features tailored specifically for AI development – such as behavior trees and the Navigation System – Unreal Engine enables creators to build sophisticated AI systems that enhance gameplay and player engagement.

If you're ready to explore the world of AI development and its potential to improve your games, then there's no better time to dive in!

Who this book is for

If you are a game programmer, or specifically an Unreal Engine developer, with little or no knowledge of video game AI systems, and want to delve deep into this topic, then this book is for you.

Developers who are proficient in other game engines and are interested in understanding the principles of the Unreal AI framework will also benefit from this book; however, a basic knowledge of Unreal Engine and C++ is strongly recommended.

A passion for gameplay logic will help you get the most out of this book.

What this book covers

Chapter 1, Getting Started with AI Game Development, gently introduces you to the realm of AI game development, starting with an understanding of the basics of AI behavior.

Chapter 2, Introducing the Unreal Engine AI System, introduces you to the main AI elements included in the Unreal Engine Gameplay Framework, such as behavior trees, the Navigation System, and the Perception System.

Chapter 3, Presenting the Unreal Engine Navigation System, introduces you to the powerful navigation capabilities within Unreal Engine, including key concepts such as navigation mesh generation and pathfinding algorithms.

Chapter 4, Setting Up a Navigation Mesh, covers essential practical techniques for implementing a navigation mesh using Unreal Engine, by starting with a concrete project.

Chapter 5, Improving Agent Navigation, introduces you to integration algorithms that will optimize the movement and interaction of AI agents within complex environments.

Chapter 6, Optimizing the Navigation System, presents some strategies and techniques to maximize the performance and efficiency of the Navigation System within Unreal Engine.

Chapter 7, Introducing Behavior Trees, introduces you to the powerful and versatile behavior tree system within the Unreal Engine framework.

Chapter 8, Setting Up a Behavior Tree, guides you through the essential steps of creating and configuring a behavior tree to drive an AI agent within Unreal Engine.

Chapter 9, Extending Behavior Trees, provides an in-depth exploration of advanced techniques for extending the capabilities of behavior trees to create more sophisticated AI behaviors and interactions.

Chapter 10, Improving Agents with the Perception System, shows how to leverage the power of the Unreal Engine Perception System to enhance the responsiveness of AI agents within virtual environments.

Chapter 11, Understanding the Environment Query System, provides a comprehensive and detailed explanation of the Environment Query System within the Unreal Engine framework.

Chapter 12, Using Hierarchical State Machines with State Trees, introduces you to the StateTree system for implementing hierarchical state machines within Unreal Engine.

Chapter 13, Implementing Data-Oriented Calculations with Mass, introduces the MassEntity framework, through which you will be able to implement efficient and scalable data-oriented calculations.

Chapter 14, Implementing Interactable Elements with Smart Objects, introduces smart objects and shows how to integrate them in an Unreal Engine environment.

Appendix - Understanding C++ in Unreal Engine, delves into the fundamental concepts and principles of using the C++ programming language within the Unreal Engine framework.

To get the most out of this book

To get the most out of this book, it is strongly recommended to have a good understanding of Unreal Engine and its main features. Some experience with C++ programming will also be an advantage. A strong passion for gaming – in particular, gameplay logic – will help you a lot in understanding the most advanced topics.

Software/hardware covered in the book	Operating system requirements
Unreal Engine 5.4	Windows, macOS, or Linux
Visual Studio 2019 or 2022 and JetBrains Rider 2023+	

As this book is focused on AI programming and not on graphics, you won't need a high-spec computer to follow all the chapters. However, to properly run Unreal Engine, a good PC with a good graphics card is highly recommended.

If you are using the digital version of this book, we advise you to type the code yourself or access the code from the book's GitHub repository (a link is available in the next section). Doing so will help you avoid any potential errors related to the copying and pasting of code.

Download the example code files

You can download the example code files for this book from GitHub at `https://github.com/PacktPublishing/Artificial-Intelligence-in-Unreal-Engine-5`. If there's an update to the code, it will be updated in the GitHub repository.

We also have other code bundles from our rich catalog of books and videos available at `https://github.com/PacktPublishing/`. Check them out!

Conventions used

There are a number of text conventions used throughout this book.

`Code in text`: Indicates code words in text, database table names, folder names, filenames, file extensions, pathnames, dummy URLs, user input, and Twitter handles. Here is an example: "Once the project is open, please check what's inside the `Content` folder."

A block of code is set as follows:

```
#pragma once

UENUM(BlueprintType)
enum class EBatteryStatus : uint8
{
    EBS_Empty = 0 UMETA(DisplayName = "Empty"),
    EBS_Low = 1 UMETA(DisplayName = "Low"),
    EBS_Medium = 2 UMETA(DisplayName = "Medium"),
    EBS_Full = 3 UMETA(DisplayName = "Full")
};
```

Any command-line input or output is written as follows:

```
$ mkdir css
$ cd css
```

Bold: Indicates a new term, an important word, or words that you see onscreen. For instance, words in menus or dialog boxes appear in **bold**. Here is an example: " In the **Details** panel, look for the **Tags** property in the **Actor | Advanced** category and hit the + button to create a new tag."

> **Tips or important notes**
> Appear like this.

Get in touch

Feedback from our readers is always welcome.

General feedback: If you have questions about any aspect of this book, email us at customercare@packtpub.com and mention the book title in the subject of your message.

Errata: Although we have taken every care to ensure the accuracy of our content, mistakes do happen. If you have found a mistake in this book, we would be grateful if you would report this to us. Please visit www.packtpub.com/support/errata and fill in the form.

Piracy: If you come across any illegal copies of our works in any form on the internet, we would be grateful if you would provide us with the location address or website name. Please contact us at copyright@packt.com with a link to the material.

If you are interested in becoming an author: If there is a topic that you have expertise in and you are interested in either writing or contributing to a book, please visit authors.packtpub.com.

Subscribe to Game Dev Assembly Newsletter!

We are excited to introduce Game Dev Assembly, our brand-new newsletter dedicated to everything game development. Whether you're a programmer, designer, artist, animator, or studio lead, you'll get exclusive insights, industry trends, and expert tips to help you build better games and grow your skills. Sign up today and become part of a growing community of creators, innovators, and game changers. `https://packt.link/gamedev-newsletter`

Scan the QR code to join instantly!

Share Your Thoughts

Once you've read *Artificial Intelligence in Unreal Engine 5*, we'd love to hear your thoughts! Scan the QR code below to go straight to the Amazon review page for this book and share your feedback.

`https://packt.link/r/1-836-20585-6`

Your review is important to us and the tech community and will help us make sure we're delivering excellent quality content.

Free Benefits with Your Book

This book comes with free benefits to support your learning. Activate them now for instant access (see the "*How to Unlock*" section for instructions).

Here's a quick overview of what you can instantly unlock with your purchase:

PDF and ePub Copies

Next-Gen Web-Based Reader

Access a DRM-free PDF copy of this book to read anywhere, on any device.

Use a DRM-free ePub version with your favorite e-reader.

Multi-device progress sync: Pick up where you left off, on any device.

Highlighting and notetaking: Capture ideas and turn reading into lasting knowledge.

Bookmarking: Save and revisit key sections whenever you need them.

Dark mode: Reduce eye strain by switching to dark or sepia themes

How to Unlock

Scan the QR code (or go to packtpub.com/unlock). Search for this book by name, confirm the edition, and then follow the steps on the page.

Note: Keep your invoice handy. Purchases made directly from Packt don't require one

Part 1: Introducing Artificial Intelligence in Games

In the first part of this book, you will receive a beginner-friendly introduction to the realm of artificial intelligence (AI) development in games. Once you have a solid understanding of its key concepts, you will be ready to start implementing a project leveraging these topics.

This part includes the following chapters:

- *Chapter 1, Getting Started with AI Game Development*
- *Chapter 2, Introducing the Unreal Engine AI System*

1

Getting Started with AI Game Development

Welcome to the fascinating world of **artificial intelligence (AI)** development in **Unreal Engine**! I am thrilled that you have chosen me and my book as your guide on this sometimes intimidating journey into the realm of AI programming. Rest assured that I am committed to making this experience as easy and enjoyable as possible.

Throughout this book, you will acquire the skills required to create Unreal Engine games that involve the use of AI techniques and learn how to handle them at runtime. We will start from the basics, such as moving agents within a game level, and gradually progress to more advanced topics such as creating complex behaviors and managing multiple AI entities (even dozens or hundreds). By the end of this journey, you will be proficient in crafting formidable opponents that will challenge your players; what's more, you will possess a deep understanding of the potential pitfalls in AI development and how to avoid them.

In this chapter, I will introduce you to some basic keywords about AI development; these concepts will serve as a gentle introduction to the whole book, providing you with a foundation to delve deeper into the fascinating world of AI programming in Unreal Engine.

In this chapter, we will be covering the following topics:

- Introducing AI
- Understanding AI in game development
- Explaining AI techniques in video games

Technical requirements

I guess you are already aware that the **Unreal Engine Editor** can be quite demanding in terms of hardware prerequisites. However, there is no need to be intimidated as this book primarily focuses on game programming rather than real-time visual effects.

In this section, we will explore the hardware and software requirements necessary to follow along with this book. Additionally, we will discuss some prerequisite knowledge that will be beneficial for your journey.

Prerequisite knowledge

Before we dive into the exciting world of AI in game development, I want to kindly remind you that this book is designed for individuals who already possess some knowledge about working with Unreal Engine. Therefore, you must be already familiar with the following topics:

- **Unreal Engine**: It's essential to have a basic understanding of this software interface, tools, and workflow.

- **Game development basics**: Having a good grasp of general game development principles and terminology will greatly aid your understanding of the concepts discussed in this book.

- **Programming knowledge**: As this book focuses on game development, it is assumed that you have some programming experience. Ideally, you should be familiar at least with the Unreal Engine visual scripting system (**Blueprints**) and, to some extent, with C++.

> **Note**
>
> If you are new to Unreal Engine, I highly recommend exploring some introductory books or resources to familiarize yourself with its fundamentals. One amazing starting point is *Blueprints Visual Scripting for Unreal Engine 5* by *Marcos Romero, Packt Publishing*, which will guide you through the main features of programming in Unreal Engine with Blueprints.

In this book, and whenever possible, I will be showing you techniques by using both Blueprints and C++. If you need a gentle introduction to C++, at the end of this book, you will find a valuable appendix that delves into the intricacies of C++ programming in the context of Unreal Engine. This quick guide is also designed to provide you with some understanding of how C++ works within the Unreal Engine framework.

Hardware requirements

At the time of writing this book, Epic Games is officially recommending the following basic requirements. If your hardware meets at least these specifications, you can expect to have a pleasant experience while reading through the chapters:

- **Windows OS**:

  - **Operating system**: Windows 10 or 11 64-bit version

  - **Processor**: Quad-core Intel or AMD, 2.5 GHz or faster

- **Memory**: 8 GB RAM
- **Graphics card**: DirectX 11- or 12-compatible graphics card

- Linux:

  - **Operating system**: Ubuntu 22.04
  - **Processor**: Quad-core Intel or AMD, 2.5 GHz or faster
  - **Memory**: 32 GB RAM
  - **Video card**: NVIDIA GeForce 960 GTX or higher with the latest NVIDIA binary drivers
  - **Video RAM**: 8 GB or more

- macOS:

  - **Operating system**: Latest macOS Ventura
  - **Processor**: Quad-core Intel, 2.5 GHz
  - **Memory**: 8 GB RAM
  - **Video card**: Metal 1.2-compatible graphics card

I've written this book using the following hardware:

- **Desktop**:

  - **Operating system**: Windows 10 64-bit version
  - **Processor**: Intel Core i9 9900K
  - **Memory**: 64 GB RAM
  - **Graphics card**: NVIDIA GeForce RTX 3090ti

- **Laptop**:

  - **Operating system**: Windows 11 64-bit version
  - **Processor**: Intel Core i7 13650HX
  - **Memory**: 8 GB RAM
  - **Graphics card**: NVIDIA GeForce RTX 4060

Software requirements

This book assumes you have the **Epic Games Launcher** and **Unreal Engine 5** installed and fully working on your computer.

> **Note**
>
> At the time of writing this book, the latest version of Unreal Engine is 5.4 but you will be able to follow along with any version more recent than 5.4.

Additionally, as we will also be working with C++, you'll need an IDE supporting this language and Unreal Engine. If you already have some experience, chances are you have already installed Visual Studio 2019/2022 or JetBrains Rider; if you don't, you will need to install one of them to follow along with the C++ coding parts.

Setting up Visual Studio for Unreal Engine development in C++

Once you have Visual Studio installed, you'll need the following extra components to make it work properly with Unreal Engine:

- **C++ profiling tools**
- **C++ AddressSanitizer**
- **Windows 10 SDK**
- **Unreal Engine installer**

To include these tools, follow these steps:

1. Open **Visual Studio Installer**.
2. Select **Modify** from your Visual Studio installation, selecting the version you will be using:

Figure 1.1 – Selecting the Visual Studio Installer version

3. Once the **Modifying** modal window opens, in the top bar, make sure you are in the **Workloads** section.

4. Then, activate the **Game development with C++** option by clicking the checkmark.

5. Next, if it is closed, open **Installation details | Game development with C++ | Optional** from the right sidebar.

6. Select the following fields, as shown in *Figure 1.2*:

   - **C++ profiling tools**

   - The latest **Windows 11 SDK** version available

   - **C++ AddressSanitizer**

   - **IDE support for Unreal Engine** (optional)

   - **Unreal Engine installer**

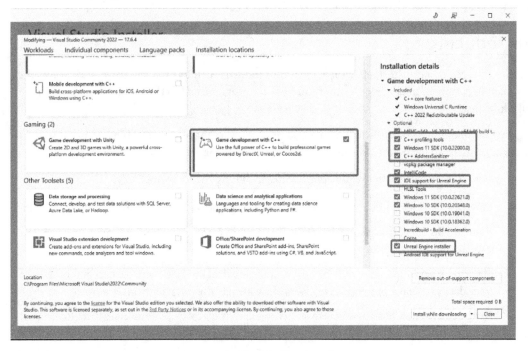

Figure 1.2 – The Workload section

7. Click the **Install while downloading** button (or the **Download all, then install** one) to start the installation process.

> **Note**
>
> The **IDE support for Unreal Engine** integration is an extension introduced in Visual Studio 2022 and adds some nifty features such as Blueprint references, Blueprint assets, and CodeLens hints on top of the Unreal Engine classes, functions, and properties. Although not mandatory, I highly recommend using it as it will make your life as a developer much easier!

After completing the download and installation process, you will be fully prepared to embark on the development of your own C++ games using Unreal Engine.

Now that you have successfully set up your system, it is time to get acquainted with some of the key terminologies in the AI environment. This will provide you with a solid foundation to understand and navigate the world of AI more effectively.

Introducing AI

AI has emerged as a transformative force in various industries; in its broadest sense, AI involves the simulation of human intelligence in machines that are programmed to think and (sometimes) learn like humans.

As such, developing AI means studying methods and software that enable machines to perceive their environment, learn from data, reason, and make decisions to achieve defined goals.

AI encompasses various subfields and applications, including the following:

- **Robotics**: The development of intelligent machines that can interact with the physical world
- **Natural language processing**: The ability of computers to understand, interpret, and generate human language
- **Machine learning**: The use of algorithms and statistical models to enable computers to learn from data and make predictions or decisions without being explicitly programmed
- **Deep learning**: A branch of machine learning that leverages neural networks to emulate decision-making abilities observed in the human brain
- **Computer vision**: The ability of computers to understand and interpret visual information from images or videos

What's more, AI has been making significant strides in the entertainment industry, transforming the way content is created, consumed, and personalized. Here are some key points about AI's impact on the entertainment industry:

- **Generative AI**: This kind of technology can create outputs such as stories, scripts, and images and has the potential to revolutionize content creation in the entertainment industry.

- **Personalized recommendations**: AI-enabled tools are being used to help users discover content tailored to their preferences by providing personalized suggestions. These recommendations are based on user behavior, viewing history, and other data, enhancing the user experience.

- **Data-driven insights**: By analyzing user behavior, AI allows the entertainment industry to gain valuable insights from data such as preferences and trends and can help companies understand their audiences better and make data-driven decisions regarding content production, distribution, and marketing.

However, as the age-old adage suggests, "*With great power comes great responsibility.*" The adoption of AI brings forth many challenges and considerations. Discussions surrounding intellectual property and copyright matters pertaining to AI-generated content have taken place and still do. Furthermore, there is growing concern about the impact of AI on employment within various industries, as certain roles may face disruption or transformation due to the advancements in AI technologies.

Ethics in utilizing AI is an essential subject that delves into the moral considerations and implications linked to the utilization of these systems. As AI technology continues to progress at a faster and faster pace, it evokes profound ethical concerns regarding its effects on society, individuals, and the environment.

While it cannot be denied that AI has the potential to enhance efficiency and productivity, it may also result in job losses within specific sectors; as such, some people consider it mandatory to implement measures that alleviate adverse effects on workers and society.

You have just been introduced to some of the most common terminology about AI in general. Now, let's shift our focus to understanding how AI works specifically in the realm of game development. In this next section, we will delve into its fundamentals, providing you with a basic understanding of its principles and workings.

Understanding AI in game development

When applied to game development, AI is employed to create intelligent systems capable of performing tasks without explicit programming. These systems adapt and improve their performance based on experience, enhancing the overall gaming experience. For instance, game characters have been imbued with AI for many years, enabling them to exhibit seemingly intelligent behavior. Even the four iconic Pac-Man ghosts have been programmed with unique and distinct behaviors!

AI in game development extends far beyond the control of **non-playable characters** (**NPCs**) or enemies. It encompasses a diverse range of applications that revolutionize game design, development, and player experience. By leveraging AI, game developers can introduce innovative and immersive gameplay elements that will captivate players during gameplay.

If you have picked up this book, chances are you are eager to grasp the fundamental principles of AI programming in games and apply this knowledge to create your next big hit. As an AI programmer, you will have the power to craft stunning opponents, create NPCs that will help players achieve their goals, or simply invent new and engaging behaviors that will make your games enjoyable to the next level; this is going to be an immensely rewarding endeavor!

However, it is important to note that AI video game programming can present significant challenges, demanding long hours and potentially inducing stress. Being aware of these potential pitfalls is crucial before embarking on this career path. To avoid such setbacks, it is essential to develop a solid understanding of how AI functions, enabling players to have a seamless and enjoyable gaming experience. What's more, comprehending this subject also entails troubleshooting computer issues that may arise and effectively resolving them. Rest assured, these issues will inevitably surface sooner or later!

In the following section, you will receive a gentle introduction to the major AI techniques used in games, along with the distinctive features that set them apart.

Explaining AI techniques in video games

AI plays a pivotal role in enhancing gaming experiences, making them more immersive and exciting. Therefore, it is crucial to have a comprehensive understanding of the underlying principles behind AI development and how they function. This knowledge will empower game developers to effectively harness AI's potential, creating rich and captivating gameplay that keeps players engaged and enthralled.

Just think about the *Assassin's Creed* series, which is known for its open-world gameplay, where complex AI behaviors are used to control NPCs. On a more advanced level, games such as *Counter-Strike* introduced AI-controlled player characters – called **bots** – that can be created and managed to stand in place of real players.

Finally, the future of AI in game development holds exciting possibilities and innovations as AI is being used to create dynamic and adaptive narratives in games. By observing player behavior and preferences, AI algorithms can construct narrative branches, challenges, and rewards tailored uniquely to each player.

In this section, I will give a brief and non-exhaustive overview of AI techniques that are commonly used in games. In this book, you'll get the chance to explore some of these techniques and see how they are used in Unreal Engine. For those techniques that won't be covered in this book, there will be plenty of opportunities for you to explore and delve into them on your own. The world of AI in gaming is vast and ever-evolving, offering endless possibilities for experimentation and innovation. So, don't be discouraged if a particular technique is not covered here – the journey of discovery continues, and there are countless resources available to help you unlock new horizons in AI game development.

Pathfinding

Pathfinding is essential for efficient navigation in game environments and refers to the process of determining the optimal path while simulating the movement from one point to another. It can be used by autonomous agents, such as NPCs or opponents, but it is also useful in point-and-click games, where your character needs to reach a specific location. Pathfinding involves finding the optimal path from one location to another while avoiding obstacles; algorithms such as **A*** are commonly used in these situations. NPCs can use this technique to plan their movement, whether to avoid enemy units, find shortcuts, or follow waypoints.

One of the most common pathfinding techniques in game development is achieved by using a **navigation mesh** – or **nav mesh**, which is a data structure that represents the walkable surfaces of a level. *Figure 1.3* shows an example of AI movement through a nav mesh:

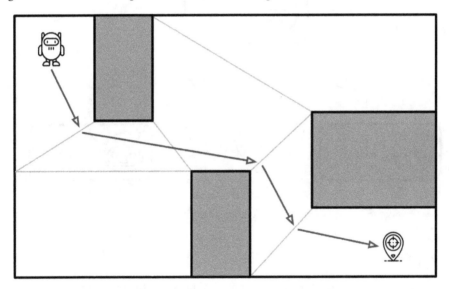

Figure 1.3 – AI movement through a navigation mesh

Rule-based

Rule-based systems refer to a type of AI that operates based on a set of predefined rules. These rules are coded by humans and dictate the behavior and decision-making of the system itself; this means following the rules to produce predetermined outcomes based on some kind of input. Put simply, these rules are commonly referred to as *if statements* because they typically adhere to the structure of *if something is true, then do something else*. Although limited, these systems are relatively easy to implement and manage because the knowledge encoded in the rules is modular, and the rules can be coded in any order. This provides much flexibility in both coding and modifying the system.

Finite state machines

Finite state machines (FSMs) are a common technique used in AI development and they involve breaking down an opponent's or NPC's behavior into different states, where each state represents a specific behavior or action. Transitions between states are triggered when certain conditions or events are satisfied. For instance, a sentinel character may have states such as patrol, alert, or chase, with transitions occurring when the character has made some noise or has been spotted because they are in the line of sight. FSMs provide a clear and organized way to control NPC behavior, especially in games with predefined sequences of actions.

Figure 1.4 shows an example of a simple FSM with states and conditions:

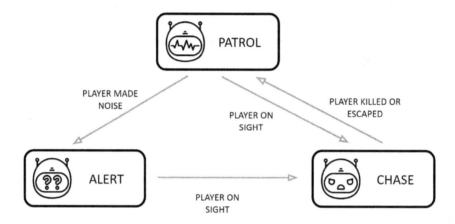

Figure 1.4 – A finite state machine

Behavior trees

Behavior trees are hierarchical structures used to control AI behavior. They consist of nodes representing specific actions or conditions. The tree structure allows for the sequencing of actions and decision-making based on the conditions themselves. The system will traverse the tree from the root to the leaf nodes, executing actions or evaluating conditions along the way. Behavior trees provide a flexible and modular approach to NPC behavior, allowing for complex and dynamic decision-making. A behavior tree can include nodes such as selectors, sequences, conditions, or action nodes. *Figure 1.5* shows a behavior tree where a selector decides which part of the tree will execute, and sequence nodes will perform a list of tasks in a predefined order.

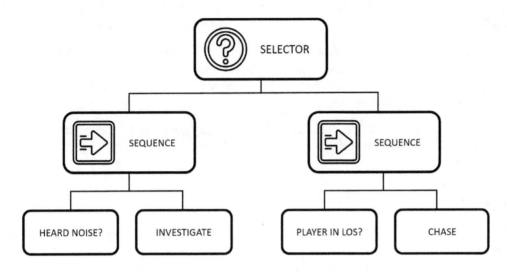

Figure 1.5 – A behavior tree

If you are unfamiliar with these terms, have no fear! I'll be explaining them in *Chapter 7, Introducing Behavior Trees*.

Machine learning AI

Machine learning involves training AI models using data and algorithms to enable NPCs to learn and improve their behavior over time. This technique allows NPCs to adapt, make decisions, and respond to unpredictable situations based on patterns and experiences from previous gameplay. Machine learning can provide more dynamic, realistic, and engaging interactions with NPCs, as their behavior evolves through iterations and learning from player actions.

One example of a game that uses machine learning is *AlphaGo* (`https://deepmind.google/technologies/alphago/`), developed by DeepMind. *AlphaGo* is an AI system that mastered the ancient Chinese game of *Go* by using machine learning techniques to calculate probabilities and make strategic decisions in the game.

Reinforcement learning

Reinforcement learning is a type of machine learning system where NPCs learn through trial and error, receiving feedback or rewards based on their actions. NPCs explore the game environment, take action, and learn from the consequences. Reinforcement learning enables NPCs to optimize their behavior by maximizing rewards and minimizing penalties. This technique can result in NPCs that exhibit adaptive and strategic decision-making, enhancing the challenge and immersion of the game. Reinforcement learning is typically employed during the development process to create a functional system by the time a game is released. Due to the nature of reinforcement learning, sometimes results may not be as expected, and NPCs may exhibit weird or erratic behavior.

Generative AI

The aforementioned generative AI is being increasingly used in video game development, offering new possibilities and transforming various aspects of game development. Some of these aspects involve creating more realistic NPCs whose behaviors go beyond fixed patterns and making decision-making systems more adaptive and engaging for players.

Although generative AI is still at its initial stages in game development and its full potential is still to be explored, it is already demonstrating promising capabilities to transform various aspects of the game industry.

Summary

In this chapter, we explored the fundamental principles of AI development and saw how it is applied in the game industry. In the upcoming chapter, I'll introduce you to the incredible potential that Unreal Engine offers and how its framework can be leveraged to create intelligent and immersive AI in games. Brace yourself for a thrilling exploration of its possibilities and let's dive into the fascinating realm of AI game programming with your favorite game engine!

Credits

The examples in this chapter were created with the help of *Basic Miscellany Lineal* icons from *Flaticon* (`https://www.flaticon.com/`).

Get This Book's PDF Version and Exclusive Extras

UNLOCK NOW

Scan the QR code (or go to `packtpub.com/unlock`). Search for this book by name, confirm the edition, and then follow the steps on the page.

Note: Keep your invoice handy. Purchases made directly from Packt don't require an invoice.

2

Introducing the Unreal Engine AI System

Welcome to the exciting world of AI programming with Unreal Engine! In this chapter, I'll be introducing you to Unreal Engine's powerful tools that will bring life and intelligence to your virtual worlds. By exploring various aspects of the Unreal Engine AI system, such as moving agents using the **Navigation System**, implementing semi-intelligent behaviors through **behavior trees** and **Blackboards**, and incorporating features such as smart objects and **mass entities**, you will gain a comprehensive understanding of the remarkable capabilities offered by this robust framework.

Mastering these skills will elevate you to the ranks of elite game programmers – and who wouldn't want to be one of those?

By the end of this chapter, you will have a sharp vision of what can be accomplished using the Unreal Engine AI system, empowering you to create advanced AI pawns in your projects.

In this chapter, we will be covering the following topics:

- Getting to know the Unreal Engine Gameplay Framework
- Presenting the Unreal Engine AI system
- Understanding advanced AI features

Technical requirements

There are no technical requirements to follow for this chapter.

Getting to know the Unreal Engine Gameplay Framework

As you may already know, Unreal Engine provides an out-of-the-box system called **Gameplay Framework (GF)** that includes many features necessary for developing a game; this spans from having an advanced input system to common entry points that will allow you to easily access data or game state.

Here are some key points explaining why the GF is so important:

- **Structure and organization**: The GF provides a structured and organized approach to developing games. It offers a collection of systems, classes, and interfaces that work together to create the core structure of a game.

- **Game logic and progression**: This framework includes predefined concepts that help define the logic, progression, and organization of a game.

- **Player and AI control**: The GF includes systems for handling player input and decision-making for characters within the game world. This encompasses player and AI control, which are essential for creating interactive and immersive gaming experiences.

- **Utility functions**: The framework provides a library of utility functions that assist with common gameplay operations and interactions. These functions can streamline gameplay logic and enhance efficiency in implementing various functionalities.

- **Flexibility and integration**: The GF is highly flexible and integrates deeply with the Unreal Engine. It uses common game programming patterns and performs heavy lifting, allowing developers to focus on building their games rather than creating their own game framework.

As a personal reflection, I have found that using and comprehending the GF over the years has significantly enhanced my overall understanding of game programming best practices.

Quite obviously, managing an AI system is also part of the GF job, so, in the next subsections, I will provide you with a concise introduction to the key AI features available in the GF, enabling you to be prepared for their use.

Actors and components

I'm pretty sure you're already familiar with **actors** and **components** in Unreal Engine, but just in case, let's do a quick refresher on both of them.

In Unreal Engine, an `Actor` class refers to any entity that can be placed within a level, whether it's a camera, a static mesh, or the player's character. An actor can undergo transformations such as translation, rotation, and scaling.

Actors serve as containers for specialized classes known as **components** that play various roles in controlling a movement, rendering, and more. There are three types of components that serve different purposes within an actor:

- **Actor components**: These primarily contain code logic for an actor. They handle various functionalities and interactions without any visual representation.

- **Scene components**: These are used to position and orient other components within the actor. They serve as reference points for transformations such as translation, rotation, and scaling but do not have any visible presence and are mainly used for organizational purposes.

- **Primitive components**: These are responsible for the visual representation of an actor within a level. They can be rendered and interacted with by players or other objects.

By combining these components, a game developer can create complex and interactive actors with both functional and visual aspects.

Main GF elements

The Unreal Engine GF is a comprehensive collection of classes that serves as a modular foundation for constructing gameplay experiences. Within this framework, game developers have the freedom to handpick specific elements that best suit the game, while being assured that these classes are intricately designed to seamlessly work together and enhance one another.

In the upcoming subsections, we will present the main elements involved to have a clear view of how things work.

GameInstance

The `GameInstance` class serves as a manager that operates behind the scenes (i.e., it is not an Unreal Engine actor); a single instance is created when the engine launches and the instance remains active until the engine shuts down. Its primary purpose is to track data and execute code as needed.

A game instance provides a handy central hub for managing persistent data, such as save game systems, and acts as a manager for other subsystems, offering convenient control over the flow of your game.

GameMode

Different from the `GameInstance` class, the `GameModeBase` or its direct descendant, `GameMode`, instance only exists in a single level and is created right after the level itself has been loaded and the world has been constructed. This class serves as a manager to handle a gameplay session, and each level can have its own different game mode logic. Its main role is to create the remaining framework actors.

GameState and PlayerState

`GameState` and `PlayerState` are specialized actors that play a key role in tracking the state of the game and the players involved. The game state is responsible for storing and handling data pertinent to all players in a game, while the player state focuses on a specific player. Given their inherent characteristics, these classes find their primary application in multiplayer games, regardless of whether they are played online or locally.

Pawn and Character

A **pawn** refers to the base class of all actors that can be controlled by players or AI entities within the game world. It serves as the physical representation of an entity, handling its involvement within the game world, including collisions and other physical interactions. It is also usually used to determine the visual appearance of an entity.

The `Pawn` class gains additional functionality through the more advanced `Character` class. The character class is specifically designed to represent players in a vertically oriented manner, enabling them to perform a wide range of actions such as walking, running, jumping, and swimming within a level. As a side note, the character class incorporates essential features for multiplayer handling.

Controller

The `Controller` class is responsible for governing the logic that determines a player's actions within the game world. Two widely used types of controller classes are `PlayerController` and `AIController`; the second option is something we eagerly anticipate, for obvious reasons.

The player controller class acts as a managerial entity, capable of processing input from a human player, enabling interaction with the game environment and facilitating their overall gameplay experience. On the other hand, the AI controller governs the actions of an AI entity by using behavior trees, state trees, navigation, and more.

The player controller and the AI controller classes can manage a character or a pawn by possessing them at runtime.

GameplayStatics

Unreal Engine provides a really helpful function library called `GameplayStatics`, which provides various utility functions for gameplay-related tasks. These functions can be used to perform common gameplay operations and interactions within the engine.

Some examples of these functions are spawning and destroying actors, retrieving information about the game world, managing gameplay tags, manipulating game instances, and more. These functions can be accessed and used from both Blueprint visual scripting and C++ programming and can streamline gameplay logic and serve as a valuable tool for managing and manipulating game elements during runtime.

Now that I have dished out some Unreal Engine GF knowledge, get ready for the juiciest part (at least in the context of this book): how AI dives into the intricate workings of the engine, equipping you with the knowledge to embark on the marvelous journey of crafting your own game logics!

Presenting the Unreal Engine AI system

Given the power of the previously described framework at your disposal, it should come as no surprise that Unreal Engine provides a comprehensive and robust AI system.

In this section, we will show the comprehensive array of tools available for Unreal Engine AI programmers along with a short description of their main features. To begin, let us examine the Navigation System and its functionality.

Navigation System

The Unreal Engine **Navigation System** allows for AI entities, called **agents**, to move on a level by using pathfinding algorithms.

The Navigation System will create a **nav mesh** derived from the geometry present within the level by using collisions. This mesh is subsequently divided into tiles, which are further partitioned into polygons, thereby forming a graph. Agents within the system use this graph to navigate toward their intended destinations. Polygons have a designated cost, which helps agents determine the most optimal path based on the lowest overall cost. Also, the Navigation System includes a range of components and settings that can be adjusted to modify the nav mesh generation process. These modifications can include alterations to the costs of polygons, influencing the navigation behavior of agents within the level. Finally, the system allows for the connection of non-contiguous areas within the nav mesh, such as platforms and bridges, thereby facilitating seamless navigation across these spatial elements. *Figure 2.1* shows a level available in the **Content Examples** project freely available on the Epic Games Launcher; the green area is the nav mesh and the character on the left is the AI agent.

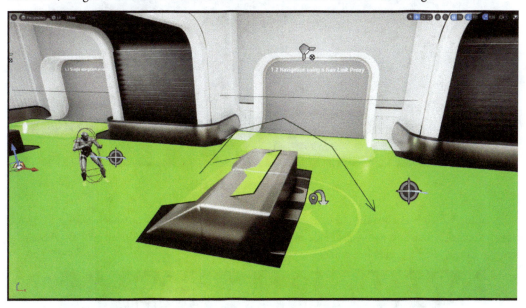

Figure 2.1 – Navigation System

Part 2 of this book will be devoted to understanding the Unreal Navigation System and how to optimize and debug it.

Behavior trees

In Unreal Engine, behavior trees serve as a valuable tool for creating AI for NPCs in your games. The primary function of a behavior tree asset is to execute branches containing logical instructions. In Unreal Engine, behavior trees are created in a pretty similar way to Blueprints – this means you will be using some kind of visual scripting method – where a sequence of nodes with specific functionality attached to them is added and connected to form a behavior tree graph.

Figure 2.2 depicts a portion of a behavior tree from the **Lyra Starter Game** project available on the Epic Games Launcher:

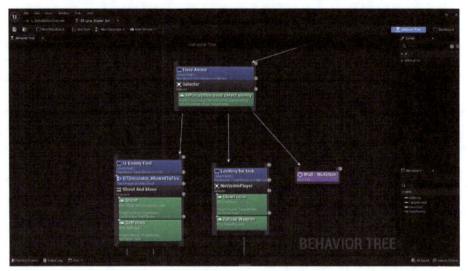

Figure 2.2 – Behavior tree example

To determine which branches should be executed, the behavior tree relies on another asset known as a Blackboard, which acts as the *brain* for the behavior tree itself. *Figure 2.3* shows the Blackboard corresponding to the previous behavior tree:

Figure 2.3 – Blackboard example

Behavior trees and Blackboards are pretty important in AI game programming; that's why I have dedicated *Part 3* of this book to this topic.

Mass Entity

The **Mass Entity** system is a gameplay-focused framework for data-oriented calculations and provides a paradigm for staging elements with behavior in the game; it is designed to handle large numbers of **entities** and facilitate behavior controls for both skeletal and static meshes.

Figure 2.4 shows a screenshot from the **City Sample** project – available on the Epic Games Launcher – that makes use of Mass Entity for both crowd and traffic control:

Figure 2.4 – Mass Entity in action

> **Note**
>
> At the time of writing this book, Mass Entity is still marked as experimental; consequently, it should be used cautiously as things may break or change as time goes by.

Mass Entity will be presented in *Part 4* of this book.

State tree

A **state tree** is a versatile hierarchical state machine that integrates some features from behavior trees with some others from state machines. With this system – organized in a tree structure – developers will be able to create highly performant logic that remains structured and adaptable. *Figure 2.5* shows a state tree from the aforementioned **City Sample** project:

Figure 2.5 – State tree example

I'll be showing you how state trees work in *Part 4* of this book.

Smart objects

In Unreal Engine, **smart objects** represent a set of activities in the level that can be used through a reservation system that ensures that only one AI agent can use a smart object at a time, preventing other agents from using it until it becomes available again. These objects are placed on a level and can be interacted with by AI agents and players. Smart objects contain all the information needed for these interactions and can be queried at runtime using dedicated filters. *Figure 2.6* shows a smart object asset from the **City Sample** project:

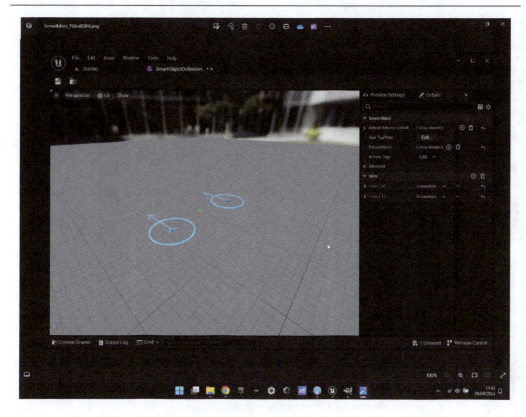

Figure 2.6 – Smart object example

Smart objects will be presented in *Part 4* of this book.

Environment Query System

The **Environment Query System** (**EQS**) collects data from the environment, enabling AIs to inquire about the data using various tests. This process results in selecting an item that best matches the question posed.

Queries can be called from a behavior tree and used to make decisions on how to proceed based on the results of the executed tests. *Figure 2.7* depicts an environment query from the **Lyra Starter Game** project.

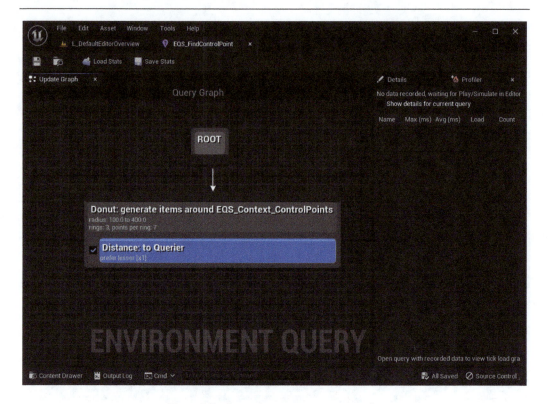

Figure 2.7 – An environment query example

> **Note**
> At the time of writing this book, the EQS is still marked as experimental so you should use it cautiously as things may break or change as development goes on.

I will be presenting you EQS by the end of *Part 4* of this book, just after you have gained a solid understanding of how behavior trees work.

AI Perception System

The **AI Perception System** provides another way for pawns to receive data from the environment, such as where noises are coming from or if the AI sees something. It allows for the generation of awareness for AI by providing sensory data for it. The system allows data sources to create stimuli so that data listeners can be periodically updated about them. This system is used to enable AI sensing within games and can react to an array of customizable sensors.

Figure 2.8 shows a character from the **Lyra Starter Game** with a stimuli source component attached:

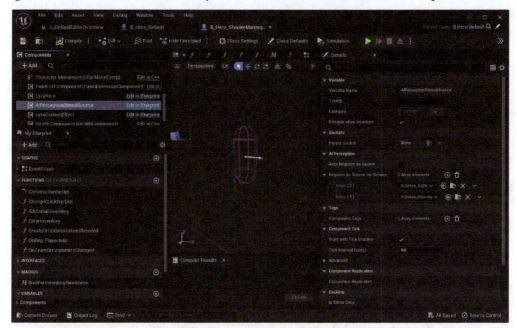

Figure 2.8 – AI perception example

AI Perception will be presented along with behavior trees in *Part 4* of this book to make your AI characters aware of their surroundings.

AI debugging

No serious framework would be complete without a debugging system. **Debugging** is an essential aspect of software development, allowing developers to identify and fix errors or bugs in their code. It plays a crucial role in ensuring the reliability and functionality of the framework.

That's why Unreal Engine offers a full arsenal of tools and features to assist developers in debugging AI, including visual debugging tools, behavior tree visualization, and AI simulation modes. These tools allow developers to inspect and modify AI behavior in real time, identify issues such as pathfinding errors or erratic decision-making, and make necessary adjustments to improve the overall AI performance within the game environment.

Figure 2.9 shows the **City Sample** project in action with the AI debugging tools enabled:

Figure 2.9 – The debugging tools enabled in a level

Throughout the book, I will be showing different techniques for debugging your game, depending on the tools you will be using. These techniques will empower you to efficiently track down and address bugs, errors, and other issues within your game's code logic.

In this section, I showed you the main AI features available in the Unreal Engine GF; in the following section, I will present some of the latest technologies that have been implemented in the engine involving **machine learning** (**ML**) systems.

Understanding advanced AI features

Now that you have a basic understanding of the main AI features available in Unreal Engine, I'd like to present to you some of the most experimental and, to some extent, non-gameplay-related features.

> **Note**
> Keep in mind that these features are still in experimental or beta release, so they need to be handled with care.

Learning Agents

Learning Agents is an experimental plugin designed specifically to enable you to train AI characters using ML. This plugin offers a unique opportunity to enhance or even replace traditional game AI systems, such as behavior trees or state machines. With learning agents, you can leverage reinforcement learning and imitation learning approaches to create intelligent and adaptive AI characters.

The primary goal of this plugin is to provide a robust solution for character decision-making in Unreal Engine. However, its potential applications extend beyond game development. As an example, Learning Agents can be used to automate testing processes by creating AI characters that perform specific actions and scenarios repeatedly. This helps identify potential issues and ensures the robustness of your game.

Although still in development, this is an impressive plugin, and you should expect more and more improvements as time goes by.

Neural network engine

The **neural network engine** (**NNE**) plugin provides developers with an API that allows unified access to different neural network inference engines. This enables programmers to seamlessly switch between inference runtimes as needed, optimizing their use case and targeting specific platforms effectively.

If you are familiar with the Unreal Engine **Rendering Hardware Interface** (**RHI**), you can think about the NNE as similar; it is a tool whose main purpose is to abstract from different inference runtimes.

ML Deformer

The **ML Deformer** is a plugin that provides an API for accessing different implementations of ML inference runtimes, allowing developers to approximate complex deformation models and improve the quality of characters' mesh deformations.

The ML Deformer is specifically designed for creating accurate non-linear deformer systems for characters in real-time game engines. It leverages some inner Unreal Engine tools to perform computations on the GPU, optimizing performance. A sample project – called **ML Sample Project** – is available on the Unreal Engine marketplace and the results are pretty amazing; *Figure 2.10* shows a lighting test made by my fellow teacher, Giovanni Visai, starting from the aforementioned sample project:

Figure 2.10 – The ML Deformer plugin in action

ML cloth simulation

The **ML cloth simulation** system offers developers a high-fidelity and highly performant solution for real-time cloth simulation. This system excels in producing clothing meshes of comparable quality to pre-simulated data while maintaining fast and efficient performance with minimal memory usage.

In conclusion, the integration of ML capabilities in Unreal Engine opens up a world of possibilities for developers. By leveraging these technologies, developers will be able to create more immersive, intelligent, and dynamic experiences within their projects.

Summary

In this chapter, I introduced you to the key features available in the Unreal Engine GF. After that, I provided an overview of the main AI systems, beginning with the Navigation System and progressing to behavior trees. Additionally, I discussed more advanced systems such as Mass Entity and state trees. Finally, I introduced you to experimental features such as the Learning Agents and NNE plugins.

Congratulations! You've reached the end of *Part 1* of this book. In the upcoming chapter, get ready to take a deep dive into the Navigation System and how to create basic AI characters that will navigate through it. So, get ready to roll up your sleeves, and let's start creating something amazing!

Part 2: Understanding the Navigation System

In the second part of this book, you will delve into the essential features of Unreal Engine's Navigation System. From there, you will create your own project and learn how to implement a fully working environment that is navigable by AI agents.

This part includes the following chapters:

- *Chapter 3, Presenting the Unreal Engine Navigation System*

- *Chapter 4, Setting Up a Navigation Mesh*

- *Chapter 5, Improving Agent Navigation*

- *Chapter 6, Optimizing the Navigation System*

3

Presenting the Unreal Engine Navigation System

The Unreal Engine **Navigation System** is a sophisticated framework that enables AI-controlled entities to navigate and interact seamlessly within game levels. It provides a set of tools and algorithms that allow game developers to define and create paths, obstacles, and movement behaviors. By using the Navigation System, you will be able to simulate realistic movement and behavior patterns of AI-controlled entities, enhancing the immersion and believability of your virtual environments. As this system incorporates advanced features such as **pathfinding algorithms**, **collision avoidance**, and **dynamic obstacle handling**, understanding its full potential is a crucial skill for aspiring AI programmers.

By the time you reach the end of this chapter, you will possess a strong comprehension of how this specific part of the Gameplay Framework operates. Equipped with this knowledge, you will be fully prepared to embark on your journey of actively working with the system itself.

In this chapter, we will be covering the following topics:

- Introducing AI movement
- Understanding pathfinding
- Testing the Navigation System with a project template

Technical requirements

There are no technical requirements to follow for this chapter.

Introducing AI movement

When it comes to moving AI entities in a virtual environment, we face numerous challenges, and there is no universal solution. The approach to solving each problem depends on each unique characteristic that will be faced in the type of game being developed. For example, is the AI's destination something stationary – for instance, a pickup – or is it something that is moving unpredictably, such as the player character? Furthermore, will the AI just need to wander around without a specific destination, or will it have a pre-defined pattern – for instance, as a patrolling sentinel?

Also, as a developer, you will need to consider factors such as different terrains, obstacles, and dangerous zones. Deciding between an easier or a more dangerous – but quicker – path can have a significant impact at runtime. These are just a few of the considerations involved in moving AIs within a level and, as you encounter different scenarios, you will be likely to face different issues. Understanding and properly addressing all pertinent variables is essential for an optimal player experience.

So, what are the main entities involved in AI movement and, specifically, in pathfinding? How do they cooperate to make a player's experience flawless? I will tell you all about it in a few seconds!

Understanding the Navigation Mesh

In Unreal Engine, the Navigation System is based on a **navigation mesh** – or **nav mesh** – that works by dividing the navigable space into polygons, which are subsequently divided into triangles for efficiency. Each triangle is then considered a node of a graph to reach a specific location and when two triangles are adjacent, their respective nodes are connected. *Figure 3.1* depicts a game level with the aforementioned mesh, divided by triangles:

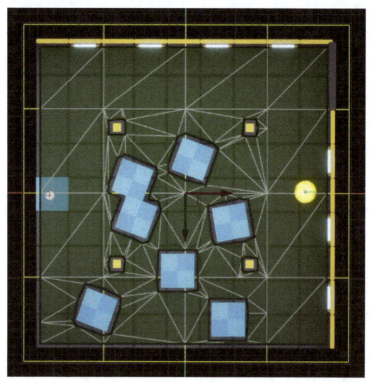

Figure 3.1 – A nav mesh example

Using this graph, you can apply any type of pathfinding algorithm – such as **A-star (A*)**, which I will explain later in this chapter – and the resulting process will generate a path among these triangles that the AI character can traverse.

Fortunately, unless you truly need to delve into the intricacies of altering the core structure of the Navigation System, there is no immediate need to venture into such detail. Understanding that the collection of generated triangles forms a cohesive graph, which serves as the foundation for pathfinding algorithms, is sufficient for getting the best out of the Navigation System.

To generate a nav mesh in Unreal Engine, all you have to do is add one or more **Nav Mesh Bounds Volume** actors in the level and change their size to suit your own needs.

> **Note**
>
> In Unreal Engine, a `Volume` class refers to a special type of actor that can influence the behavior of other actors within its area of effect. Volumes are used to define various effects, such as lighting, and can modify how players or other objects interact with the game world. Some common types of volumes in Unreal Engine include Trigger Volumes, Lightmass Importance Volumes, Post-Processing Volumes, and, obviously, Nav Mesh Bounds Volumes.

Figure 3.2 shows a **Nav Mesh Bounds Volume** actor added to a level; the yellow lines mark the volume itself.

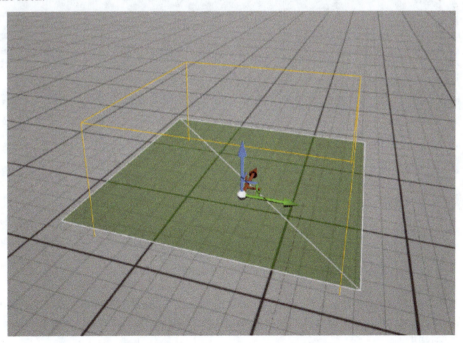

Figure 3.2 – A Nav Mesh Bounds Volume

This actor is pretty straightforward, and the only possible action you can take with it is to adjust its extension. From the previous image, you may have noticed a green mesh, made of two triangles; this fairly simple shape has been generated by another actor: the **Recast Nav Mesh** one, which is usually auto-generated the very first time Nav Mesh Bounds Volume is added to the level.

This actor oversees the walkable area generation for AI entities that will use it to make their own efficient and accurate calculations on pathfinding; an instance of it is usually auto-generated once you add a Nav Mesh Bounds Volume in your level.

It should be noted that most of the settings available for the RecastNavMesh actor can be set with predefined values in your editor's **Project Settings** – this can be opened from the **File** menu – by selecting the **Engine - Navigation Mesh** section, as depicted in *Figure 3.3*:

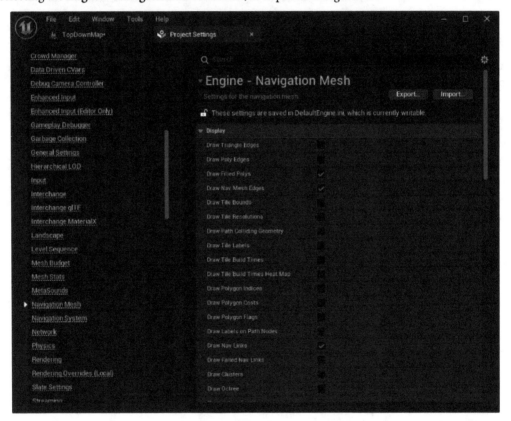

Figure 3.3 – The Navigation Mesh settings section

Now that you understand how a nav mesh is created in Unreal Engine, it is important to know that you can adjust it to enhance its interest and realism using navigation modifiers by using the available modifier system.

Modifying the nav mesh

The Navigation System comprises various actors and components that alter the generation of the nav mesh, such as the cost of traversing a polygon. These adjustments influence how AI agents move through your level.

The Nav Modifier Volume

The simplest one is the **Nav Modifier Volume** actor, whose task is… well, to modify a nav mesh! Once you have positioned this volume in your level, you will have the choice to modify how an AI agent will perceive it for pathfinding – you can designate it as an impassable terrain, a difficult terrain, or even an obstacle. *Figure 3.4* shows three modifier volumes set with the three different cost settings:

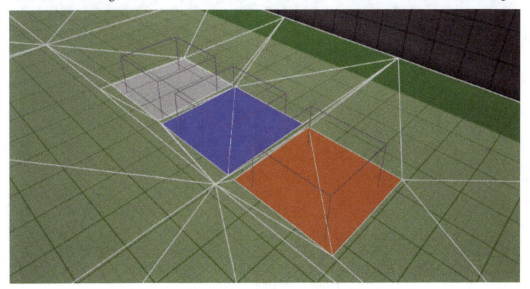

Figure 3.4 – Three navigation modifiers in action

If you are thinking about creating your own nav mesh modifiers, well, that's possible; you will just need to extend the UNavArea class, set its parameters, and you are ready to go!

Navigation Query Filters

As an additional method for tailoring the behavior of the Navigation System when generating paths for AI agents, you can take advantage of **Navigation Query Filters**. This method encompasses information pertaining to one or more specific areas and provides the flexibility to override the cost values assigned to the areas themselves, if necessary. By implementing query filters, you will gain the ability to customize the navigation patterns of AI agents as they traverse various regions within your game world, and this will let you fine-tune and optimize the movement of AI entities.

Navigation Link Proxies

When you begin designing your walkable terrain, you will most probably be introducing gaps or areas with varying altitudes; and I guess you'll need your AI character to jump from one side to the other. That's exactly why the Nav Link Proxy has been created; this actor will connect two areas of the nav mesh that lack a direct navigation path. *Figure 3.5* shows such a link, connecting two zones at different heights:

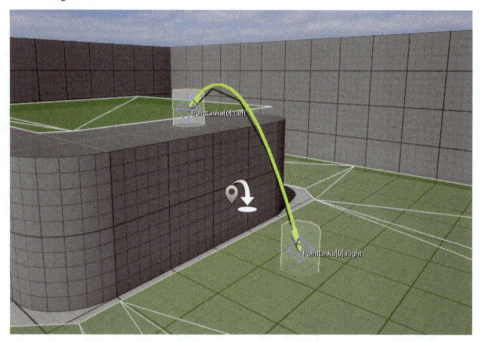

Figure 3.5 – Nav link proxy

With such a useful tool at your disposal, you will be able to make your character jump, fall down, and execute breathtaking acrobatics, seamlessly transitioning from one gravity-defying move to another.

Runtime nav mesh generation

By default, Unreal Engine is set to generate nav meshes statically – this means that the mesh is generated offline and cannot be changed at runtime. However, if you need a more flexible way of generating a nav mesh, you can opt for the **dynamic mesh generation** system that will let you update the mesh under different circumstances – for example, by adding moving entities. The runtime generation can be enabled for the whole project by opening **Project Settings**, and then going to the **Engine - Navigation Mesh** section, and selecting the **Runtime Generation** option in the **Runtime** category:

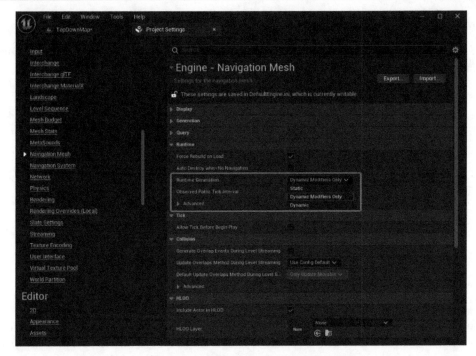

Figure 3.6 – Enabling Runtime Generation

Alternatively, you can enable it for just a single level, by changing the **Runtime Generation** attribute in the **Recast Nav Mesh** actor.

Invokers

A **Navigation Invoker** is an actor component that will generate the nav mesh around an AI agent at runtime. It is used to remove the need for pre-computing the mesh, and it allows for dynamic navigation in the game world. This feature is particularly useful when you have extensive terrains – for instance, an open world – that will take a huge amount of time to generate the nav mesh; through an Invoker, the system will generate the mesh itself at runtime, but just in a limited space, around the actor.

You now have a grasp of how a nav mesh is created and tweaked, so let's explore who – or what – will be moving through it.

AI agents

In Unreal Engine's Navigation System, an **agent** is an AI character or entity that is capable of navigating the game world by means of the nav mesh. An agent will use the nav mesh data to compute paths, avoid obstacles, and move around the environment intelligently. Each agent in a level typically represents a specific type of character, such as a player character – for instance, in a point-and-click game – an enemy AI, or any other entity that needs to move within the game world.

To move an agent within a level, you will typically be using a **Pawn** or a **Character** actor.

One of the easiest ways to move an agent toward a location or an actor is by using the **Simple Move to Location** or **Simple Move to Actor** Blueprint nodes.

Figure 3.7 – The Simple Move nodes

Alternatively, you can use the corresponding C++ methods called, respectively, `SimpleMoveToLocation()` and `SimpleMovetoActor()`.

Obviously, in Unreal Engine, you can create more complex behaviors than just moving an agent to a single point; this is something we will explore starting from *Chapter 8, Setting Up a Behavior Tree*.

Avoidance

Basic pathfinding algorithms are effective for finding routes around stationary objects; however, when it comes to moving obstacles – such as player characters or other AI agents – a more suitable system is needed. That's why Unreal Engine provides two **avoidance** systems, to prevent collision between moving entities:

- The **Reciprocal Velocity Obstacles (RVO)** system computes the velocity vectors for each agent, considering nearby agents and assuming they are moving at a constant velocity in each time step of the calculation. The chosen optimal velocity vector is the closest match to the agent's desired velocity in the direction of its destination. This system is included in the character movement component. RVO does not use the nav mesh for avoidance, so it can be used separately from the Navigation System for any character.

- The **Detour Crowd Manager** computes a rough sample of velocities that lean toward the agent's direction, resulting in a substantial enhancement in avoidance quality compared to the standard RVO approach. This system can be used by any actor extending the `Pawn` class by using the `ADetourCrowdAIController` class.

You will be introduced to avoidance in *Chapter 5, Improving Agent Navigation*.

In this section, you have been introduced to the main elements involved in pathfinding and how they interact with the environment.

In the following section, I will provide you with further details on how pathfinding works.

Understanding pathfinding

As you are aware, Unreal Engine uses **pathfinding** for moving an agent around a level; in this section, I will go a bit deeper into detail on how things work under the hood. Unreal Engine takes advantage of a generalized version of the A* algorithm, a widely employed graph traversal and pathfinding algorithm in computer science. Known for its completeness, optimality, and efficiency, its main goal is to determine the shortest path between a designated source node and a specified goal node in a weighted graph.

This graph is a node-based representation of the level, where nodes represent walkable areas that are interconnected and have information on neighbor nodes and traversal costs to reach them.

A* uses a heuristic function to estimate the cost from each node to the target location; this trial-by-error system helps guide the search toward the most promising paths, improving efficiency.

During the pathfinding process, the algorithm maintains two lists: one of them contains nodes that are yet to be evaluated, while the other contains nodes that have already been evaluated. The algorithm evaluates each node by considering its cost, and the cost of reaching it from the previous node. It selects the node with the lowest total cost from the open list for further evaluation. Once the target node is reached, the algorithm reconstructs the path by backtracking from the target node to the start node, following the connections between nodes.

Unreal Engine's version often includes post-smoothing operations to improve the quality of the generated path. Post-smoothing adjusts the path to make it more natural and avoid obstacles more effectively.

If you want to take a deep dive into how the nav mesh generation works and how pathfinding is computed, my suggestion is to check the Unreal Engine source code available on GitHub (`https://github.com/EpicGames/UnrealEngine`); in particular, you should look for the `NavigationSystem` and `NavMesh` modules, which are located in the `Engine/Source/Runtime` folder.

> **Note**
>
> To access the Unreal Engine source code, you will need to be part of the Epic Games GitHub organization. Subscription is free and there's no reason why you shouldn't take part in it.

As an example, by checking the `DetourNavMeshQuery` class in the Unreal Engine source code, you will get an insight into how the A* pathfinding algorithm is used and how cost is computed or how to find a tile on a path.

It seems you have gained some insight into how pathfinding is handled within Unreal Engine, so I guess it's time to delve into a practical example by exploring a real case scenario; we will begin by creating a project from a template.

Testing the Navigation System with a project template

In this section, we will look at a project that uses the Unreal Engine Navigation System and do it with a project template – using a pre-made project such as a template presents a valuable chance for you to acquire practical experience on a particular topic, saving you the time and effort required to build a project from the ground up.

Once the project has been created, we will quickly analyze the nav mesh generation system and how the template handles the character movement at runtime.

Here, you will start by creating a game prototype by using the **Top Down** project, one of the templates available in the **GAMES** category of **Unreal Project Browser**.

Setting up the project

Once you are ready, you can fire up the Epic Games Launcher and follow these steps:

1. Select **GAMES** | **Top Down** from the available templates.

2. Set the project to **Blueprint** or **C++** depending on your personal preferences.

3. Name your project – any name will do.

4. Leave the other settings with their default values.

5. Click the **Create** button.

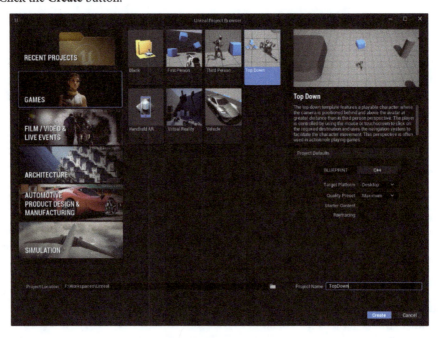

Figure 3.8 – Project setup

Once the project has been created and opened, you are ready to analyze it.

Analyzing the nav mesh

We are now going to get a brief tour of the generated level and of the actors that contribute to the nav mesh generation.

In the **Outliner** view, you will notice that there is a folder named `Navigation`, including three actors:

- **NavMesh Bounds Volume**
- **Recast Nav Mesh**
- **Nav Link Proxy**

Let's analyze each element in detail.

Nav Mesh Bounds Volume

As you already know from the previous sections in this chapter, the **NavMeshBoundsVolume** actor is responsible for defining the area where the nav mesh will be computed. By selecting it, you will notice that a yellow-edged area is shown, wrapping all the game levels, as shown in *Figure 3.9*:

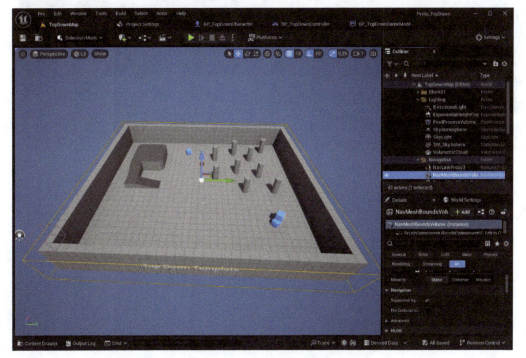

Figure 3.9 – Nav Mesh Bounds Volume

You cannot do much with this actor; just resize it and the nav mesh will be recomputed.

Recast Nav Mesh

The **RecastNavMesh** actor will take care of the nav mesh generation; by default, it does not have a visible representation within the Unreal Engine Editor. However, if you press the *P* key on your keyboard, the nav mesh will become visible and accessible within the editor interface. *Figure 3.10* shows the level once this actor has been made visible:

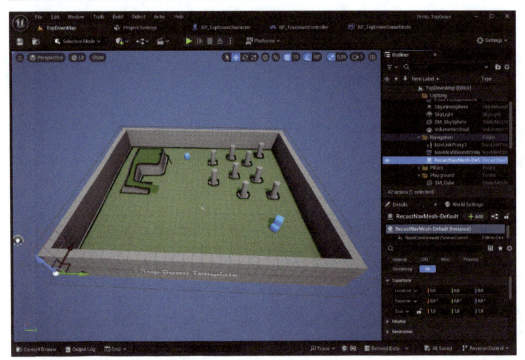

Figure 3.10 – Recast Nav Mesh

In *Chapter 6*, *Optimizing the Navigation System*, I will be presenting you some techniques on how to optimize mesh generation with this actor. For now, you can just check the **Display** category in the **Details** window; here, you will have access to a plethora of visualization utilities that will come in handy later in this book. As an example, in *Figure 3.11*, I am showing the sections of the level – named **tiles** – along with their labels and the generated polygons:

Figure 3.11 – Some display settings for the Recast Nav Mesh actor

You may have observed that the blue cube actors on the level are not affecting the nav mesh in any way. This is because they have been configured not to impact the navigation; since they are movable objects, we do not intend for them to create non-navigable areas around them.

As a simple test, in the **Details** panel, you can look for the **Can Ever Affect Navigation** property and enable it; the nav mesh will be immediately recomputed and the cube will carve a hole in it, as depicted in *Figure 3.12*:

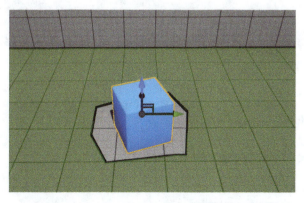

Figure 3.12 – The blue cube actor carving the nav mesh

The aforementioned property states that the object will be an obstruction in the navigation area and that the generated hole will be a non-navigable area in the mesh.

Please note that letting such a movable object carve the nav mesh may produce undesired results; by default, the nav mesh is static and cannot be altered at runtime. This implies that even if the object is moved, the non-navigable area will remain fixed and – though unseen – will hinder the player character from moving into or through it.

Nav Link Proxy

The last nav mesh entity in this example is a **NavLinkProxy** actor that, in our level, will let the player character jump down from a platform.

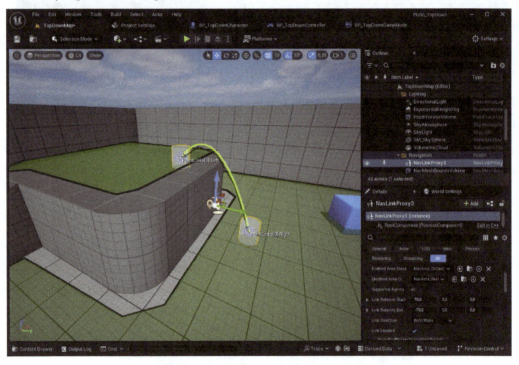

Figure 3.13 – The Nav Link Proxy in the level

Now that you comprehend how the nav mesh is structured, let's examine the character controller to understand how the player character is maneuvered.

Analyzing the character controller

Depending on your choice in creating the project – Blueprints or C++ – you will have two slightly different versions of the character controller.

The Blueprint character controller

The code for moving the controller character is pretty straightforward and can be found in BP_ TopDownController – located in the Content/TopDown/Blueprints folder.

Once the Blueprint class is opened, locate the MoveTo function in the **Functions** tab and open it; you will find the **Simple Move to Location** node that is used to make the player character move through the nav mesh.

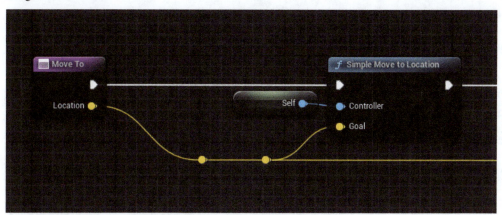

Figure 3.14 – The MoveTo function

This is all you need to use to move your player character to a predefined location in the level.

The C++ character controller

The same logic can be found inside a C++-generated project; just open the generated .cpp class for the player controller and look for the OnSetDestinationReleased() method; you will find this line of code:

```
UAIBlueprintHelperLibrary::SimpleMoveToLocation(this,
CachedDestination);
```

This helper function will start the navigation process for your agent.

Testing the project

Now that you know all the actors involved in this project, you can simply hit the **Play** button and start testing how the Navigation System works. In particular, you will notice that the character will move to the destination point by selecting the shortest path.

Additionally, once on the raised platform on the left, it will be able to jump down thanks to the Nav Link Proxy that has been added at that location.

Summary

In this chapter, I introduced you to the key components of the Unreal Engine Navigation System. We started by discussing the process of generating the nav mesh, which is essential for AI agents to navigate the environment. Then, I provided a brief explanation of how the pathfinding algorithm works, enabling AI characters to find their way efficiently. Lastly, I highlighted the benefits of using the **Top Down** project template, which effectively utilizes the Navigation System.

By now, you should have a good understanding of the capabilities offered by the Unreal Engine AI Navigation System. I imagine you are eager to dive in and start coding on your keyboard, and that is exactly what we will be doing in the next chapter!

Get This Book's PDF Version and Exclusive Extras

UNLOCK NOW

Scan the QR code (or go to `packtpub.com/unlock`). Search for this book by name, confirm the edition, and then follow the steps on the page.

Note: Keep your invoice handy. Purchases made directly from Packt don't require an invoice.

4

Setting Up a Navigation Mesh

I'm pretty confident that, by now, you are aware that a pivotal element in AI game development is establishing a fully functional nav mesh. This entity acts as the cornerstone for directing AI-controlled agents through the game world with accuracy and effectiveness.

In this chapter, we will begin to put this understanding into practice by initiating a new project. By the conclusion of the chapter, you will have hands-on experience in developing and refining a navigation system within your own project.

This knowledge will serve as a crucial milestone in shaping your path as an AI programmer, propelling you toward the creation of groundbreaking games that will revolutionize the gaming industry!

In this chapter, we will be covering the following topics:

- Introducing Unreal Agility Arena
- Creating an AI agent
- Setting up a basic level
- Adding navigation modifiers
- Working with navigation link proxies

Technical requirements

To follow along with this chapter, you should have set up Visual Studio (or JetBrains Rider) with all Unreal dependencies, as explained in *Chapter 1, Getting Started with AI Game Development*.

You'll be using some starter content that's available in this book's companion repository at `https://github.com/PacktPublishing/Artificial-Intelligence-in-Unreal-Engine-5`. Through this link, locate the section for this chapter and download the following `.zip` file: `Unreal Agility Arena - Starter Content`.

If you somehow get lost while going through this chapter, in the repository, you will also find the up-to-date project files here: `Unreal Agility Arena - Chapter 04 End`. Also, to fully understand this chapter, it is necessary to have some basic knowledge about Blueprint visual scripting while I guide you through the key characteristics of setting up an AI agent.

Introducing Unreal Agility Arena

To kick off a successful project, it's essential to have a solid foundation. Imagine diving into a short novel starting like this:

In a secret underground lab hidden beneath a nondescript building, an eccentric scientist named Dr. Markus toiled away on his latest invention: AI dummy puppets. These puppets were no ordinary puppets; they were equipped with advanced AI technology that made them capable of interacting with the environment in the most unexpected ways.

Dr. Markus was known for his quirky personality and wild ideas. He believed that these puppets held the key to understanding human behavior and improving social interactions. With his trusty sidekick, Professor Viktoria, by his side, he embarked on a series of hilarious experiments.

Well, it looks like you've stumbled upon the perfect starting point for creating the next big hit in the video game industry, and your task is to craft mind-blowingly awesome AI agents that will rock the world of gaming by seamlessly interacting with their virtual surroundings!

Explaining the project brief

The project you'll be working on will be a set of **gym** levels, where you will be creating different behaviors for your AI agents.

> **Note**
> In game development, a gym typically refers to a training environment where developers can test and train their AI algorithms and models. This term is also commonly used in the context of reinforcement learning, where AI agents learn to play games through trial and error in simulated environments. For the purpose of this book, we will stick to the first definition.

In my personal opinion, working on a gym is one of the most entertaining parts of the prototyping phase of a game because you don't need to worry a lot about things working perfectly; you can experiment with all sort of things, and – in the end – you will most probably come up with creative and unconventional solutions!

So, to start with, I have provided a project template – called Unreal Agility Arena – that you will be using during the rest of this book. After downloading and opening it in Unreal Engine, our main focus will be on creating self-contained levels to experiment with the knowledge acquired thus far. This will involve addressing small tasks and resolving them effectively.

Once you reach *Chapter 7*, *Introducing Behavior Trees*, you will be ready for something more challenging, and things will get a bit tougher, but also – I promise you – much more engaging and interesting!

> **Note**
>
> As this book is about AI game programming rather than game design, balancing game mechanics will not be a primary focus of gameplay. Instead, the focus will be on making things work effectively.

The very first step involves cracking open the project and diving into the delicious assets I've served up for you. So, let's get started!

Starting the project

While the project will mainly focus on gym levels, we aim for a visually appealing, or – as we game developers like to say – *juicy* look and feel. I understand that many of you may not have a background in 3D modeling (and neither do I!). That's why we will be using some fantastic assets by Kay Lousberg (`https://kaylousberg.com/`) that are available for personal and commercial purposes.

Figure 4.1 – Kay Lousberg's website

> **Note**
>
> In this project, I have mainly used the `Prototypes Bits Extra` package that is freely distributable once you have bought a license. If you are thinking about using the models for a commercial project, please consider buying a license from Kay's website as well.

After downloading the file from the link provided at the start of this chapter, unzip it and open the project by double-clicking the `UnrealAgilityArena.uproject` file.

Once the project is open, please check what's inside the `Content` folder. You will see these subfolders:

- A `_GENERATED` folder that has some additional models I have created with the Unreal Engine **modeling tools**
- A `KayKit` folder that includes all models from Kay
- A `Maps` folder that includes some pre-made levels that are ready for use
- A `Materials` folder that contains some materials needed by the project assets
- A `Textures` folder that contains some textures used by the project materials
- A `Vfx` folder that contains some Niagara effects that we will be using later

With all these resources at our disposal, we are prepared to begin creating elements for the project, starting with an agent that will navigate through our level.

Creating an AI agent

As the first step in testing the pathfinding system, we are going to create an agent whose sole aim is to reach a target actor inside the level; it won't be anything fancy, just an actor that will be able to reach a target point in the level.

Let's start by creating a new folder inside **Content Drawer** and calling it `Blueprints`. Double-click on the newly created folder to open it and perform the following steps:

1. Right-click on **Content Drawer** and, from the menu that opens, select **Blueprint Class**.

2. From the **Pick Parent Class** window that will open, select **Character**, as depicted in *Figure 4.2*:

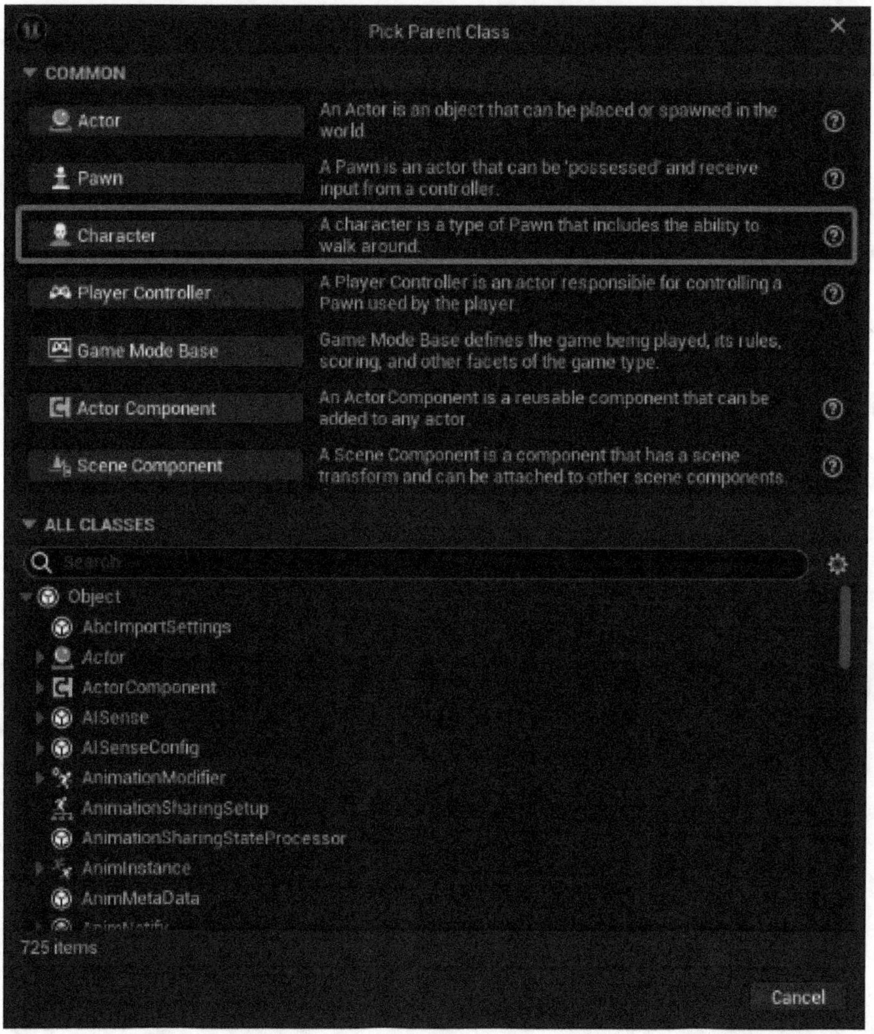

Figure 4.2 – Character creation

3. Name the newly created asset **BP_NavMeshAgent**.

4. Double-click on it to open it.

As you probably already know, the `Character` class refers to a specific type of pawn that is designed to represent players or AI agents in a vertically oriented manner, allowing them to walk, jump, fly, and swim through the game world.

We will start by giving it a visual representation and setting up the main values.

Creating the agent

With the `Blueprint` class opened and the **Viewport** tab selected, locate the **Details** panel and follow these steps:

1. Open the **Skeletal Mesh Asset** property dropdown and select the **Dummy** asset.

2. Open the **Anim Class** property dropdown and select the **ABP_Dummy** asset.

3. In the **Character Movement (Rotation Settings)** category, locate the **Max Walk Speed** property and set the value to `500.0 cm/s`.

4. In the **Character Movement (Rotation Settings)** category, locate the **Rotation Rate** property and set the **Z** value to `640.0°`.

5. In the same category, check the **Orient Rotation to Movement** checkbox.

6. In the **Shape** category, set the **Capsule Half Height** property to `120.0` and the **Capsule Radius** property to `50.0`.

7. In the **Pawn** category, uncheck the **Use Controller Rotation Yaw** property.

The previous steps are quite straightforward, and they will just set up the character mesh, assign an animation Blueprint – that has already been created for you – and, finally, set the capsule component size and the movement rotation settings. The final result for the agent is shown in *Figure 4.3*:

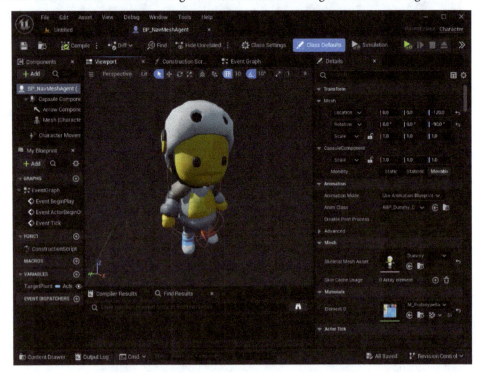

Figure 4.3 – The agent Blueprint

The agent is almost ready, we just need to add some simple code logic in order to make it reach a target in the level.

Adding the navigation logic

Open the **Event Graph** tab and follow these steps:

1. Add a **Get AIController** node in the graph and connect its incoming **Controlled Actor** pin to a **Reference to Self** node.

2. Click and drag from the outgoing **Return Value** pin of the **Get AIController** node and add a **Move to Actor** node to the graph.

3. Connect the outgoing execution pin of the **Event BeginPlay** node to the incoming execution pin of the **Move to Actor** node.

4. From the incoming **Goal** pin of the **Move to Actor**, click and drag and, once released, select the **Promote to variable** option; name the newly created variable `TargetActor`.

5. In the **My Blueprint** tab, select the **TargetActor** variable and, in its **Detail** panel, check the **Instance Editable** property.

   This is pretty simple; at the start of the game, the agent will try to navigate to a target actor; setting a variable to **Instance Editable** will make it visible in the level, in order to pick up the agent destination. The visual scripting code you have just created is shown in *Figure 4.4*:

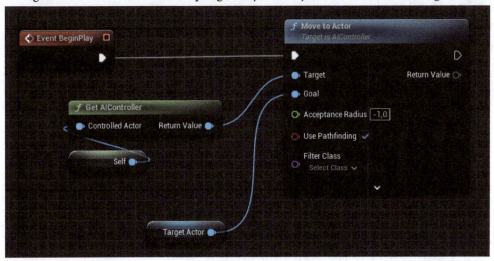

Figure 4.4 – The Blueprint graph

Our maybe not-so-clever agent is now fully equipped to navigate toward its designated target position. In the upcoming section, we will create a suitable environment for the agent to freely move around and explore.

Setting up a basic level

We are now going to create our first level and start testing the pathfinding system for the agent. The project template has some prefabs I created for you, particularly a set of **Packed Level Actors** for quickly prototyping your maps and a **Level Instance** for setting up the lighting system. You are more than welcome to create your own game levels, but during this phase, my advice is to follow along with what I will be doing.

> **Note**
>
> In Unreal Engine, a Level Instance allows you to create reusable instances of a level or a portion of a level; this way, you can efficiently duplicate and reuse parts of your level design without having to recreate them from scratch. A Packed Level Actor is a type of Level Instance that is optimized for rendering and can only contain static meshes. Level Instances and Packed Level Actors are particularly useful when you have complex or repetitive elements in your level that you want to reuse multiple times.

Creating the level

To create our first gym, follow these steps:

1. From the main menu, select **File | New Level**.

2. Navigate to the `Maps/LevelInstances` folder and drag an instance of **LI_Lighting** inside your level; set its transform **Location** to (0, 0, 0).

3. Navigate to the `Maps/PackedLevelActors` folder and drag an instance of **PLA_Lab_01** inside your level; set its transform **Location** to (0, 0, 0).

4. From the `KayKit/PrototypeBits/Models` folder, drag some obstacles into the level just to make things a bit more engaging for your agent; my level is shown in *Figure 4.5*:

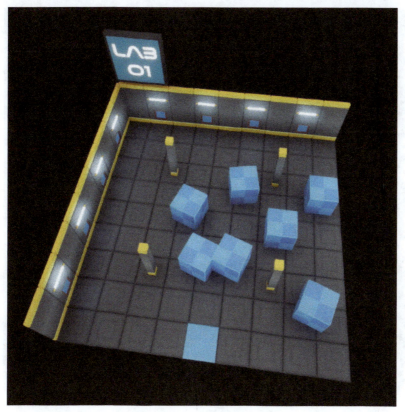

Figure 4.5 – The base level

5. Save the level in the `Maps` folder and name it `Gym_NavMesh_01`.

Adding the nav mesh

You are now ready to add the nav mesh on the level:

1. From the **Quickly add to the project** button in the toolbar, select **Nav Mesh Bounds Volume** and drag an instance in the level; a **Recast Nav Mesh** actor will be automatically added along with the volume.

2. Set **Location** for the volume to (`0, 0, 0`) and **Scale** to (`20, 20, 1`).

3. Click inside the level and hit the *P* key on your keyboard to show the generated nav mesh, as depicted in *Figure 4.6*:

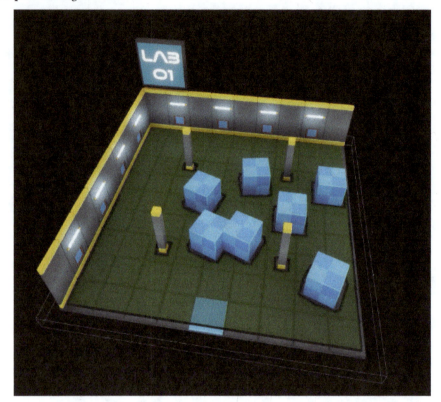

Figure 4.6 – The nav mesh

As you can see, the obstacles you have added will carve the nav mesh and make things more interesting for the soon-to-be-added agent.

Adding the agent

As a last step, we need to add the agent and a target point to be reached. So, let's start by doing this:

1. From the Vfx folder, drag an instance of the **NS_Target** Niagara system and place it anywhere on the nav mesh.

2. From the Blueprints folder, drag an instance of the **BP_NavMeshAgent** Blueprint and place it on the blue-colored tile of the level; the **Location** value should be approximately (-1650, 30, 180).

3. With the agent selected, locate the **Target Actor** property in the **Details** panel and, from the dropdown menu, set its value to **NS_Target**, which is the previously added Niagara system. The final level should be similar to *Figure 4.7*:

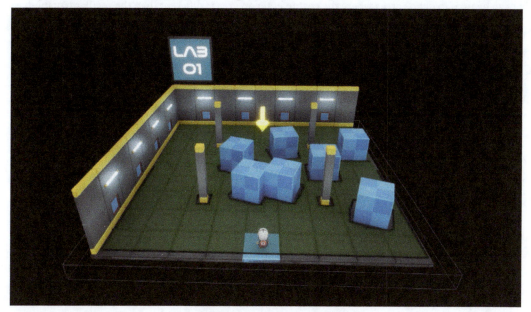

Figure 4.7 – The final level

Now that the level is complete, we can start testing it out.

Testing the gym

Now that the gym is finished, you can start testing your agent to see how it behaves in the level.

You can simply hit the **Play** button in the toolbar, or you can use the **Simulate** button that will not enter play mode but will show a simulation of how the level works. In this context, I personally prefer this second option as it will keep the nav mesh visible.

Once the simulation starts, you will see the agent reaching the target actor by taking the shortest route possible. You are free to experiment with obstacles to check how the agent behaves in different scenarios.

In this section, you have begun to experiment hands-on with how the pathfinding system operates. You achieved this by creating an agent and a simple gym environment for your agent to navigate.

In the next section, you will be adding modifiers to the nav mesh to give your agent a bit more of a challenge.

Adding navigation modifiers

In this section, we are going to create another gym that will let us test nav mesh modifiers – actors that can be used to define areas where the cost to enter an area is different than the regular nav mesh.

We'll start by creating the level and then by adding the modifiers.

Creating the level

To create our second gym, follow these steps:

1. From the main menu, select **File** | **New Level**.

2. Navigate to the `Maps/LevelInstances` folder and drag an instance of **LI_Lighting** inside your level; set its transform **Location** to (0, 0, 0).

3. Navigate to the `Maps/PackedLevelActors` folder and drag an instance of **PLA_Lab_04** inside your level; set its transform **Location** to (0, 0, 0).

4. Save the level in the `Maps` folder and name it `Gym_NavMesh_02`.

    Now, repeat the same steps you have done for the previous gym by doing the following:

5. Add the **Nav Mesh Bounds Volume** actor and set its boundaries so they cover the full walkable area.

6. Add the **NS_Target** Niagara System on the opposite side of the level in relation to the blue tile.

7. Add the **BP_NavMeshAgent** Blueprint on the blue tile and set the **Target Actor** property value to **NS_Target**. The level should now look like *Figure 4.8*:

Figure 4.8 – The updated nav mesh

So far, everything is quite similar to the previous gym; we are now going to insert a modifier and see how it behaves.

Incorporating a modifier

We are now ready to change the way the pathfinding system behaves by adding a modifier. To do so, follow these steps:

1. From the **Quickly add to the project** button in the toolbar, select **Nav Modifier Volume** and drag an instance in the level.

2. Set **Location** for the volume to (0, 0, 0) and **Scale** to (5, 20, 1).

   You will notice that the nav mesh has now been modified and that it has been carved where the brown – let's say muddy – tiles are placed.

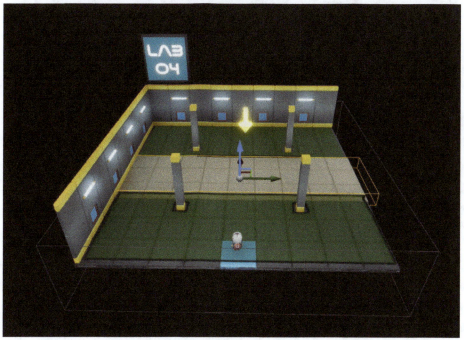

Figure 4.9 – The level with the modifier added to it

If you test the level now, you will see the agent moving toward the target point but stopping next to the muddy terrain; this is happening because the modifier volume has changed the nav mesh and now there's no way for the agent to reach its target.

By selecting the **Nav Modifier Volume** actor and checking the **Area Class** attribute in the **Details** panel, you will notice that it has been set to a value equal to `NavArea_Null`. This value applies an infinite cost to the area it is applied to, making it impossible for the agent to traverse it.

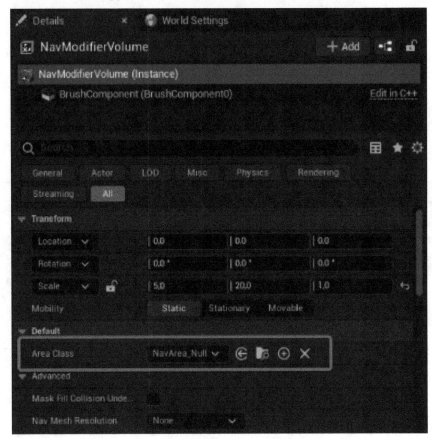

Figure 4.10 – The Area Class attribute

If you try setting this value to `NavArea_Default`, you will notice that the nav mesh will behave as it would without any modifier in it, and that's exactly what it does for this value; the cost for traversing this section is the same as that for the regular mesh. By testing the gym now, you will notice that the agent will walk through the muddy terrain and get to the target point.

If you need to check the cost for traversing each polygon in your nav mesh, you can do it by selecting the **Recast Nav Mesh** actor and – from the **Details** panel – checking the **Draw Polygon Cost** attribute. *Figure 4.11* shows the cost visualization in the level:

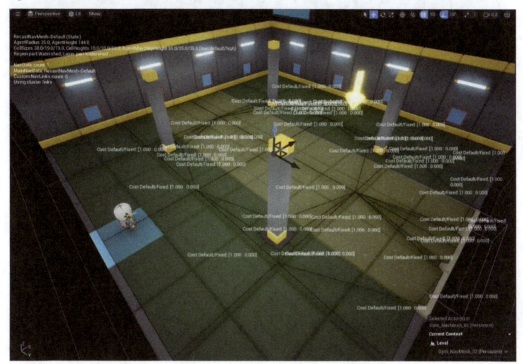

Figure 4.11 – The nav mesh traversal costs

Let's make the level a bit more interesting and add some custom-made modifier volumes.

Improving the level

Let's now create a safe path for our agent so it doesn't get its feet dirty in the muddy area. We will create a passage through it by using some additional models. To do this, follow these steps:

1. In the KayKit/PrototypeBits/Models folder, locate the **Pallet_Large_Pallet_Large** model.

2. Drag three instances of this model in the level in order to create a passage through the muddy area, as depicted in *Figure 4.12*:

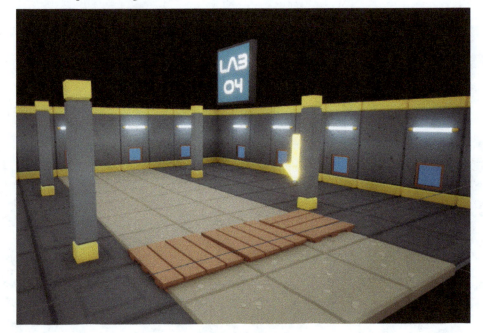

Figure 4.12 – The bridge

It's time to add another modifier for the bridge:

1. Modify the size of the previously added **Nav Modifier Volume** actor in the level so that it covers the muddy passage on the left of the river.

2. Add another **Nav Modifier Volume** actor to the level so that it covers the muddy passage on the right of the river.

3. Add a third **Nav Modifier Volume** actor to the level and place it so it creates an area covering the whole bridge.

In Unreal Engine, a **Nav Modifier Volume** is an actor that is used to change the way the nav mesh is generated and that can be added to the level to specify certain areas of the nav mesh itself.

You should now have a totally impassable zone, created by three modifiers, as shown in *Figure 4.13*:

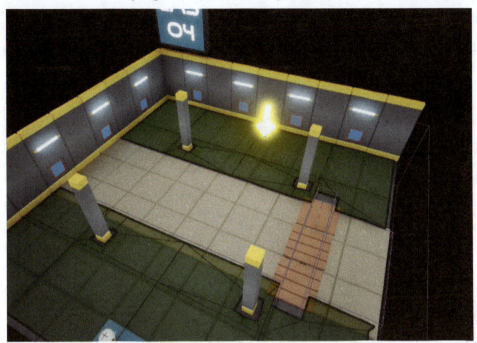

Figure 4.13 – The bridge with the modifiers

Now that we have added a safe passage, we are going to create custom modifiers in order to experiment with different settings.

Creating custom modifiers

A modifier is a Blueprint extending the `NavArea` class; this means that you can create your own modifiers by subclassing this type and setting your own parameters.

We are now going to create one for the muddy surface and one for the bridge. Let's start with the first one:

1. Open **Content Drawer** and, in the `Blueprints` folder, right-click and choose **Blueprint Class**.

2. From the **ALL CLASSES** dropdown list, select the **NavArea** type.

3. Name the newly created asset `NavArea_Mud` and double-click on it to open it.

4. Change its values as follows:

    *   Change **Default Cost** to `10.0`.

    *   Change **Fixed Area Entering Cost** to `2.0`.

- Change **Draw Color** to a recognizable color of your choice.

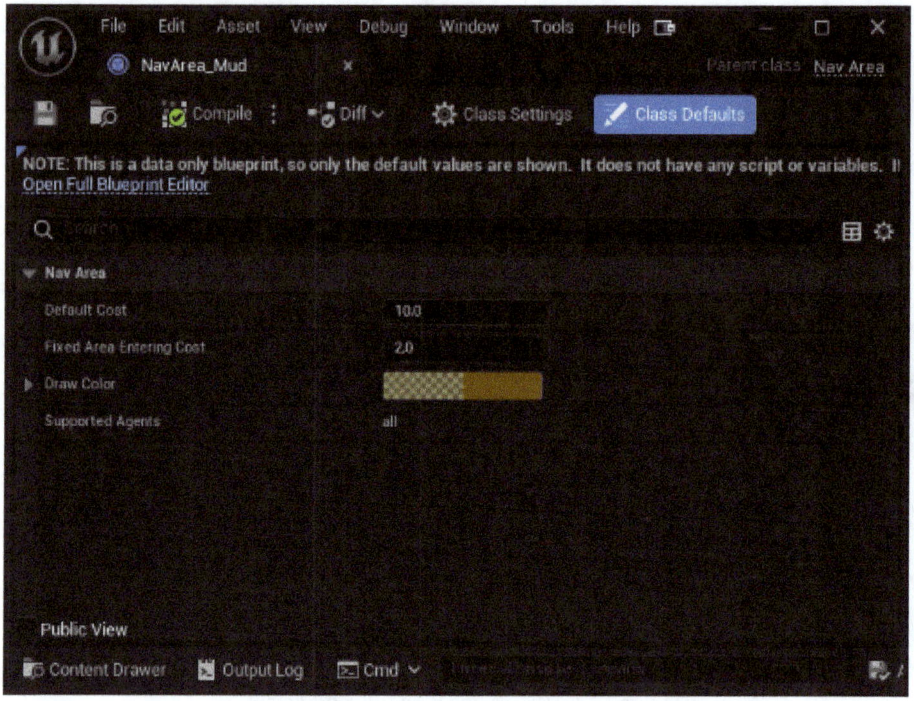

Figure 4.14 – The cost settings for the muddy nav area

While the **Draw Color** value is almost self-explanatory – it will be used to color the nav mesh area covered by the volume – **Default Cost** is a multiplier applied to the overall cost for traversing the area. This means that the pathfinding system will compute the traversal cost multiplied by the value of **Default Cost**. **Fixed Area Entering Cost**, on the other hand, is a cost that is applied only once – when the agent enters the area covered by the volume. In this case, we have opted for a fixed cost for entering the muddy area and a high cost for traversing it.

Now, let's do the same steps for the bridge area:

1. Open **Content Drawer** and, in the `Blueprints` folder, right-click and choose **Blueprint Class**.

2. From the **ALL CLASSES** dropdown list, select the **NavArea** type.

3. Name the newly created asset `NavArea_Bridge` and double-click on it to open it.

4. Change its values as follows:

   - Change **Default Cost** to `5.0`.

   - Change **Fixed Area Entering Cost** to `0.0`.

- Change **Draw Color** to a recognizable color of your choice.

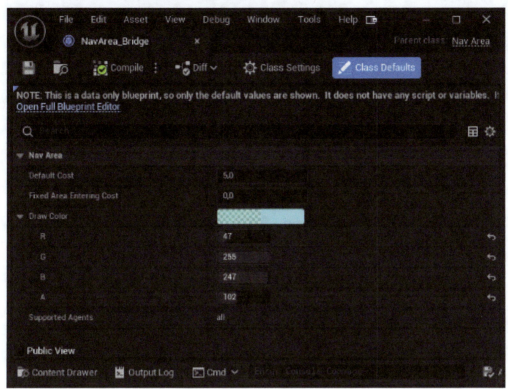

Figure 4.15 – The cost settings for the bridge nav area

In this case, we want to create an easier path for the agent, so we have set the fixed cost to 0.0 and the traversal cost to a lower value.

We now need to apply these classes to the modifiers in the level.

Applying the custom modifiers

We are now ready to get the newly created classes and apply them to the level modifiers. To do so, follow these steps:

1. Select the two mud **Nav Modifier Volumes** and, in the **Area Class** property dropdown of the **Details** panel, select **NavArea_Mud**.

2. Select the bridge **Nav Modifier Volume** and, in the **Area Class** property dropdown of the **Details** panel, select **NavArea_Bridge**.

The nav mesh should be updated and should look like the one shown in *Figure 4.16*:

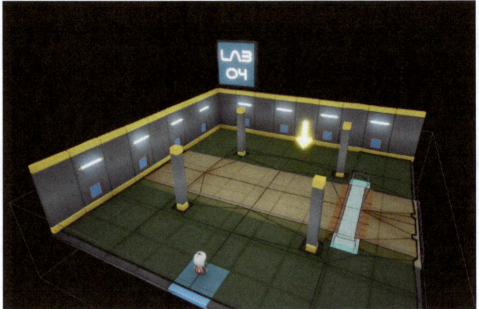

Figure 4.16 – The modified nav mesh area

> **Note**
>
> Depending on how you created your bridge, you may have a less regular modifier area than mine; additionally, the nav mesh may be generated so that it is not fully walkable. To solve these issues, you will need to play a bit with the pallet model's **Z** value and the size of your modifier.

We are finally ready to test this level.

Testing the level

To test your gym, simply start the level simulation; you should see your agent going toward the bridge, crossing it, and reaching the target point. Although the muddy terrain is walkable, passing through it has a higher cost than traversing the bridge.

To double-check it, add some impassable obstacles on the bridge, just like I did in *Figure 4.17*:

Figure 4.17 – The obstructed bridge

If you start the simulation, you will see the agent going straight to the target point; although the muddy terrain has a high traversal cost, there is no other viable solution so the agent will opt for it.

This concludes this section, where you learned about how modifiers work. In the next section, I will show you another method to modify your nav meshes using link proxies.

Working with navigation link proxies

As we have already seen in *Chapter 3, Presenting the Unreal Engine Navigation System*, a **Nav Link Proxy** is an actor used to define specific areas where agents can navigate even if a portion of the level cannot be traversed. A **Nav Link Proxy** is placed in the game world to mark a start point and an endpoint, creating a navigation link. This link will provide a connection – which can be mono- or bi-directional – between two areas that may not be directly accessible.

To check how this link works, we'll be creating a new gym.

Creating the level

To create this new gym, follow these steps:

1. From the main menu, select **File | New Level**.

2. Navigate to the `Maps/LevelInstances` folder and drag an instance of **LI_Lighting** inside your level; set its transform **Location** to (`0, 0, 0`).

3. Navigate to the `Maps/PackedLevelActors` folder and drag an instance of **PLA_Lab_03** inside your level; set its transform **Location** to (`0, 0, 0`).

4. Save the level in the `Maps` folder and name it `Gym_NavMesh_03`.

This gym has a wide water channel in the middle of it, along with a bridge, as depicted in *Figure 4.18*:

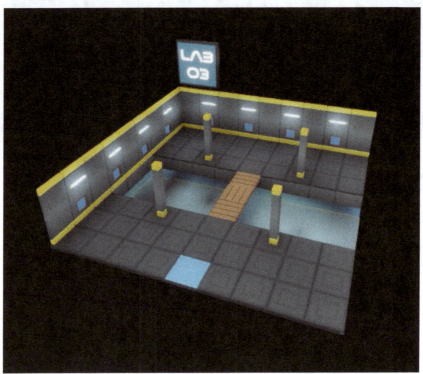

Figure 4.18 – The gym

Now, let's add a few nav meshes – one for each walkable area – along with the agent and the target point:

1. Add two **Nav Mesh Bounds Volume** actors and set their boundaries so that they cover the walkable areas on each side of the river.

2. Add the **NS_Target** Niagara System on the opposite side of the level from the blue tile.

3. Add the **BP_NavMeshAgent** Blueprint on the blue tile and set the **Target Actor** property value to NS_Target. The level should now look like *Figure 4.19*:

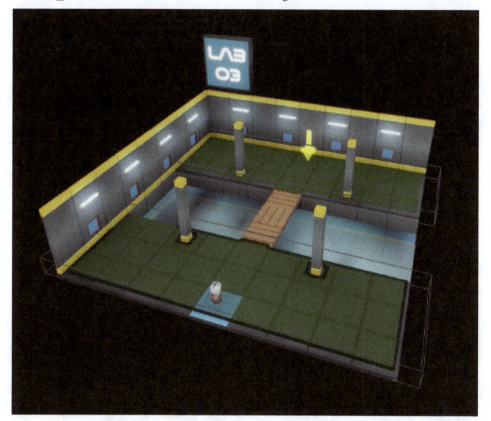

Figure 4.19 – The gym with the nav mesh

If you test your gym right now, you will notice your agent walking toward the target point but stopping near the channel. Well, I guess that's about as unexpected as a penguin wearing a tuxedo to a fancy party – there is no connection, so there can be no successful pathfinding!

Let's now allow the agent to walk through the bridge.

Adding a Nav Link Proxy

To add a link that will connect the two sides of the channel, do the following steps:

1. From the **Quickly add to the project** button in the toolbar, select **Nav Link Proxy** and drag an instance in the level.

You will notice that this actor has a couple of diamond-shaped gizmos called, respectively, **PointLinks[0].Left** and **PointLinks[0].Right**; those are the connection points that will shape your link proxy.

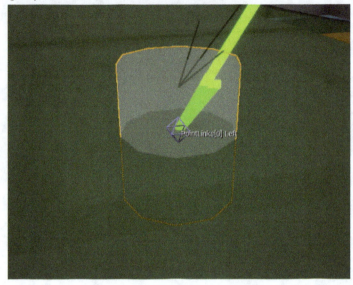

Figure 4.20 – One of the two point links

2. Select each of the point links and move them so that they are placed on each side of the bridge, as depicted in *Figure 4.21*:

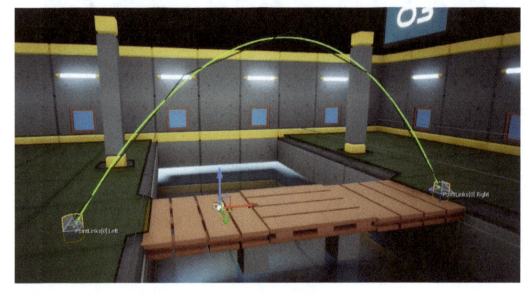

Figure 4.21 – A nav mesh proxy

The **Point Links** attribute is a list of elements that can be used to create connections between unreachable locations. By default, there is just one, but you can add as many as you want.

Testing the gym

Now that also this gym is finished, you can test it to see how your agent behaves. Once the simulation starts, you will see the agent reach the target point by traversing the bridge as if a nav mesh existed on it.

> **Note**
>
> If you try placing some obstacles on the bridge, you will notice your agent running straight on them and stopping; this happens because there is no actual nav mesh on the bridge, but a straight link. This means that Nav Link Proxies are a powerful tool, but they need to be used wisely.

In this section, you have seen how to use Nav Link Proxies to connect unreachable sections of your nav mesh; with this new knowledge, you're all set to make your AI friends do some epic cliff dives and pull off all sorts of wacky tricks!

Summary

In this chapter, you started working with the Navigation System; by starting from a simple navigable area, you added some obstacles and checked how the agent behaves. Next, you learned how to modify your walkable areas by adding non-traversable zones or difficult terrain. Lastly, you gained some understanding of how to link disconnected parts of the nav mesh.

All this knowledge is crucial because it helps your AI agents figure out where they can go without bumping into walls or getting stuck in a maze of confusion. You can think of the nav mesh as the GPS for your digital pals, ensuring they don't end up lost in the virtual wilderness!

In the next chapter, we will get deeper into the realm of the nav mesh; get ready to create some more advanced and fascinating stuff in Unreal Engine!

5

Improving Agent Navigation

Now that you have a strong grip on the basics of the Unreal Engine pathfinding system, it's time to get deeper and start delving into the intricate workings of enhancing the agent Navigation System.

In this chapter, you will discover how to improve your nav mesh generation and agent movement: starting from dynamically generated meshes, going through querying the environment, up until avoiding other agents efficiently.

By the end of this chapter, you will have some brand-new skills to make your levels more engaging and interesting. This knowledge will serve as a fundamental building block in guiding your path toward creating sophisticated games, ultimately enhancing your skills as an AI game programmer.

In this chapter, we will be covering the following topics:

- Generating navigation meshes at runtime
- Influencing navigation with query filters
- Implementing agent avoidance

Technical requirements

To follow the topics presented in this chapter, you should have completed the previous ones and understood their content.

Additionally, if you would prefer to begin with code from the companion repository for this book, you can download the `.zip` project files provided in this book's companion project repository at `https://github.com/PacktPublishing/Artificial-Intelligence-in-Unreal-Engine-5`.

You can download the files corresponding to the end of the last chapter by clicking on the `Unreal Agility Arena - Chapter 04 - End` link.

Generating navigation meshes at runtime

Let's continue our short novel started in *Chapter 4, Setting Up a Navigation Mesh*:

In the secret research laboratory, a groundbreaking experiment was underway. The artificial intelligence dummy puppets, developed by Dr. Markus and Professor Viktoria, were primed for a quest: equipped with advanced pathfinding systems, the puppets were cleverly placed in the lab's complex network of corridors and interconnected rooms.

They faced simulated construction zones, unexpected barriers, and even simulated distractions that obstructed their path. However, armed with their state-of-the-art AI capabilities, the puppets swiftly adapted to the changing circumstances, employing their ingenuity to find alternative routes and skillfully navigate through the intricate layout of the laboratory.

If you are working with nav meshes, chances are you will sooner or later have objects that will be moving around and will cause the agent's path to the target to change.

This is why having a static nav mesh generation will become useless; you will need some kind of system that will update the nav mesh at runtime. This is why Unreal Engine provides more than one method for generating such meshes.

As explained in *Chapter 3, Presenting the Unreal Engine Navigation System*, the generation method can be changed from **Project Settings** or the **Recast Nav Mesh** actor in a level; as we need to change the nav mesh generation just for this level, we will be selecting the second option.

But first, we need to create a level for our puppet agent to walk on!

Creating the level

To create this gym, do the following steps:

1. From the main menu, select **File | New Level**.
2. Navigate to the `Maps/LevelInstances` folder and drag an instance of **LI_Lighting** inside your level; set its transform **Location** value to (0, 0, 0).
3. Navigate to the `Maps/PackedLevelActors` folder and drag an instance of **PLA_Lab_02** inside your level; set its transform **Location** value to (0, 0, 0).
4. Save the level in the `Maps` folder and name it `Gym_NavMesh_04`.

This gym has a block of stairs along with a detached platform, as depicted in *Figure 5.1*:

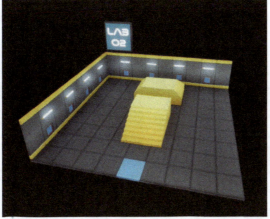

Figure 5.1 – The gym

Now, let's add the nav mesh along with the agent and the target point:

1. Add a **Nav Mesh Bounds Volume** actor and set its boundaries so it covers the whole level.

2. Add the **NS_Target** Niagara system at the top of the isolated platform.

3. Add the **BP_NavMeshAgent** Blueprint on the blue tile and set the **Target Actor** property value to NS_Target. The level should now look as follows:

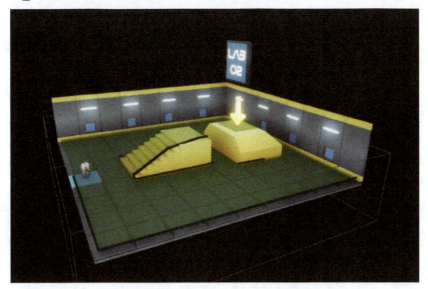

Figure 5.2 – The gym with the nav mesh

As we are using a level that is slightly more complex than those presented in the previous chapter, you may experience a mesh generation that is slightly different from the one depicted in my screenshot; the stairs nav mesh may seem detached from the rest, or you may experience some other issue. *Figure 5.3* shows a typical scenario you may encounter:

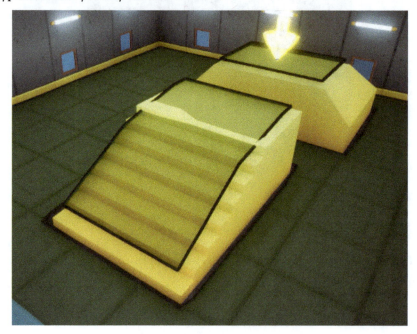

Figure 5.3 – Broken nav mesh

To solve this type of problem, you will need to tweak the **Recast Nav Mesh** actor settings a bit. As an example, you may try to do the following:

1. Set the **Default Cell Height** property value to 40.0.
2. Set **Agent Radius** to 70.0.

The first value will make the stairs navigable area connect to the rest of the level, while the second one will add some more padding to the border of the mesh so as to avoid an agent walking too close to the edges. Just play around with the values a bit until you get your desired result.

If you test the level right now, you will notice that the agent will try to reach the target point but it will get stuck at the base of the detached platform; obviously, there isn't a route to the destination, and your little agent may not make it all the way there, but it will give its best shot to get as close as it can.

Now, try to add some obstacles around the level to force your agent to climb the stairs – something like *Figure 5.4*:

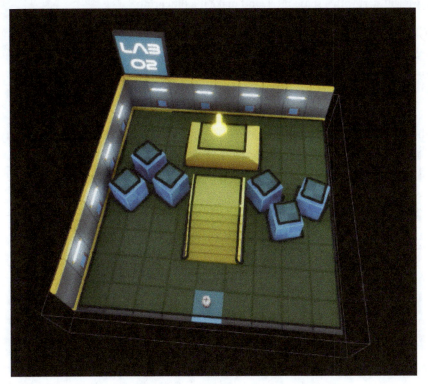

Figure 5.4 – The gym with some obstacles

If you test the level, the agent will stop at the very first step of the stairs; this is happening because the step height is too high for the poor fellow! To solve this, open the **BP_NavMeshAgent** Blueprint, and in the **Details** panel, look for the **Character Movement: Walking** category and set the **Max Step Height** value to 55.0.

This will allow the agent to take higher steps and climb the stairs. Test the level again and you should see the agent climbing the stairs and stopping at the very edge of them. The journey to the ultimate destination appears to be quite a challenge for our little friend!

Let's help it out by adding a pathway to the isolated platform, where the target point is located.

Adding a moving platform

To make our agent reach its target point, we are going to add a moving platform; this will allow us to create a dynamically generated nav mesh.

Creating the Blueprint

To do this, we will start with a static mesh and make it a Blueprint:

1. In **Content Drawer**, open the KayKit/ProtorypeBits/Models folder and drag a **Primitive_Cube_Primitive_Cube** instance into the level.

2. Set its **Location** attribute to (380.0, 50.0, 60.0) so that it connects the staircase and the isolated yellow platforms.

3. Set **Mobility** of this actor to **Movable**.

4. With this actor selected, convert it into a Blueprint by clicking the **Convert to Blueprint** button.

Figure 5.5 – The Convert to Blueprint button

5. Save the Blueprint in the Blueprints folder and name it BP_MovingPlatform.

6. Open the Blueprint by double-clicking the asset.

We are now going to add some code logic, in order to make it move.

Adding the code

With the **BP_MovingPlatform** Blueprint open, start doing the following steps:

1. Create a new variable of the Vector type and name it StartLocation.

2. Create a new variable of the Float type and name it VerticalOffset. From the **Details** panel, check the **Instance Editable** property.

   These two variables will store the initial position of the platform and the vertical distance it will cover while moving, respectively. The second one has also been made editable from the level.

Now, in **Event Graph**, perform the following steps:

3. Add a **Set Start Location** node and connect its incoming execution pin to the outgoing execution pin of the **Event Begin Play** node.

4. Add a **Get Actor Location** node and connect its **Return Value** outgoing pin to the **Start Location** incoming pin of the previously added setter node.

5. From the outgoing execution pin of the **Set Start Location** node, add a **Timeline** node and double-click on it to start editing.

This part of the code stores the platform's initial position and initializes a timeline to set up the platform animation. The code so far is shown in *Figure 5.6*:

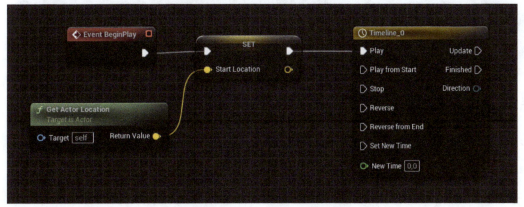

Figure 5.6 – The first part of the graph

The timeline is going to be a simple sinusoidal curve that will control the vertical offset of the platform. To create the curve, perform the following steps:

6. Click the **+ Track** button and, from the dropdown menu, select **Add Float Track**.

7. Name the track `Alpha`.

8. Set the **Length** value of the curve to `15.00`.

9. Click the **Loop** button to make the **Timeline** repeat indefinitely.

10. In the curve graph, add three keys by right-clicking and selecting the **Add key** option. Set the keys' values, respectively, to the following:

    - **Time** to `0.0` and **Value** to `0.0`.

    - **Time** to `7.5` and **Value** to `1.0`.

    - **Time** to `15.0` and **Value** to `0.0`.

11. Right-click on each key and set the **KEY INTERPOLATION** value to **Auto**; this will make the curve sinusoidal and not linear. The resulting curve is shown in *Figure 5.7*:

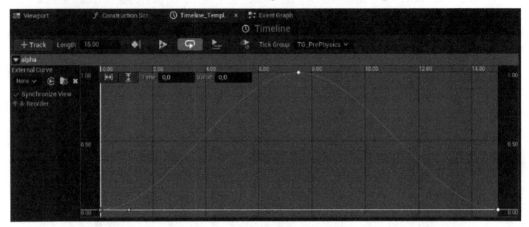

Figure 5.7 – Timeline curve

Get back to **Event Graph** and you should notice that the **Timeline** node now has an **Alpha** outgoing pin; we will use it to control the vertical position of the platform. Start doing the following steps:

12. Add a **Get Vertical Offset** node.

13. Add a **Get Start Location** node; right-click on its outgoing pin and select **Split Struct Pin** to expose the **X**, **Y**, and **Z** pins of the structure.

14. Add a **Multiply** node and connect its two incoming pins to the **Vertical Offset** outgoing pin and to the **Alpha** outgoing pin of **Timeline**.

15. Add an **Add** node and connect its two incoming pins to the outgoing pin of the **Multiply** node and the outgoing **Start Location Z** pin of the **Get Start Location** node.

16. Add a **Set Actor Location** node to the graph, connect its incoming execution pin to the outgoing execution pin of the **Timeline** node, and do the following:

    I.   Right-click on the **New Location** incoming pin and select **Split Struct Pin** to expose the **X**, **Y**, and **Z** pins of the structure.

    II. Connect the incoming **New Location X** pin to the outgoing **Start Location X** pin of the **Get Start Location** node.

    III. Connect the incoming **New Location Y** pin to the outgoing **Start Location Y** pin of the **Get Start Location** node.

IV.   Connect the incoming **New Location Z** pin to the outgoing pin of the **Add** node. The result of this part of the graph is shown in *Figure 5.8*:

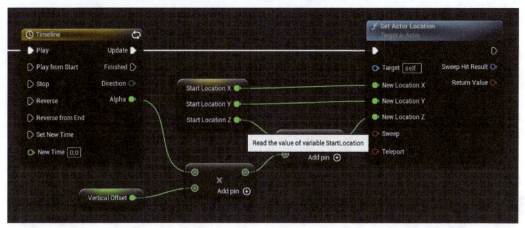

Figure 5.8 – The second part of the graph

Although quite lengthy, this code is pretty straightforward; it simply uses the **Timeline** node to compute an offset over time and applies it to the platform's **Z** position. At runtime, the platform will keep going up and down, creating a passage to the target point once every 15 seconds – that is, the duration of the loop.

If you simulate the gym right now – with the nav mesh visible – something weird will happen; although the platform will be moving up and down, the nav mesh will stay as originally generated, leaving a floating passage.

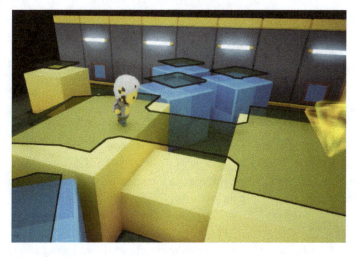

Figure 5.9 – Static generation

This issue is happening because the nav mesh is generated statically and won't be changed at runtime; the poor little fellow will drop off while trying to reach its target point, believing that a passage exists!

Let's fix this immediately by making the nav mesh dynamic.

Making the nav mesh dynamic

As stated at the beginning of this section, we are going to set the mesh runtime generation active just for this level, so we need to change the **Recast Nav Mesh** actor. To do this, select the **Recast Nav Mesh** actor and, in the **Details** panel, locate the **Runtime** category. Set the **Runtime Generation** dropdown value to **Dynamic**.

If you test the gym, you will now see the nav mesh updating at regular intervals and the passage will be interrupted.

Figure 5.10 – Dynamic generation

You may notice that the agent will stop at the edge of the stairs platform and won't go on when the moving platform creates the passage. To fix this small issue, do the following:

1. Open **BP_NavMeshAgent** and disconnect the **Event BeginPlay** node.

2. Connect the **Move to Actor** incoming execution pin to the outgoing execution pin of the **Event Tick** node.

3. Open the **Class Defaults** tab and set the **Tick Interval (secs)** attribute to 0.5, to make the update a bit sparser. The updated Blueprint is shown in *Figure 5.11*.

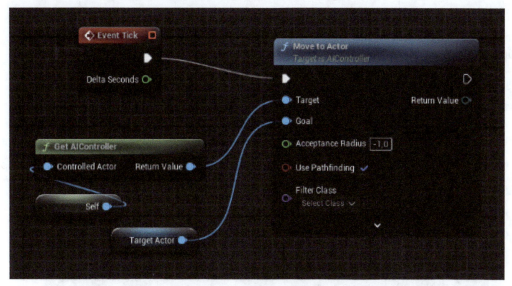

Figure 5.11 – Updated Blueprint

What we have done here is pretty simple; instead of executing the code once at the start of the game, we set it to be executed once every tick – set to half a second – in order to keep on setting the final destination. This may not be the most performance-efficient solution, but it should work well for our small prototype.

> **Note**
>
> Keep in mind that updating a nav mesh at runtime is heavy on computation; this means that you should use this feature only when necessary and stick to the static generation most of the time.

Now that you have mastered creating dynamic nav meshes, your tiny little agent will be in for a challenge as it navigates toward its target point. Life – or rather, artificial life – won't be as simple for it anymore!

In the next section, I'm going to introduce you to navigation filters, a feature that will let you change the way an agent behaves on the nav mesh.

Influencing navigation with query filters

As already mentioned in *Chapter 3, Presenting the Unreal Engine Navigation System*, with query filters, you can custom tailor the navigation paths of AI agents, enabling you to enhance and optimize their movement in the environment.

You may have noticed that the **Move to Actor** method in the previously created graph has an attribute named **Filter Class** (see *Figure 5.11*); this will allow us to customize the way our agent behaves on a nav mesh.

A filter is created by extending the `NavigationQueryFilter` class and setting some appropriate values, so let's start by creating one such class and see how it works.

Creating the level

As a first step, we will need a gym for testing filters; in this case, we are going to duplicate a pre-existing one and tweak it a bit:

1. Duplicate the **Gym_NavMesh_02** map, rename it `Gym_NavMesh_05`, and open it.

2. Remove all the obstacles from the bridge – if any.

3. Duplicate the agent and put it at the side of the first one, on the blue tile.

4. To make the agents recognizable, you may wish to change the material of the duplicate; in my case, I opted for **MI_Prototype_B**. The final level should be like the one depicted in *Figure 5.12*:

Figure 5.12 – The base gym

If you test the level, unsurprisingly, both agents will go toward the bridge, traverse it, and get to the target point.

Let's spice things up a bit, shall we?

Creating the query filter class

We are now going to create a query filter that will override the way navigation modifiers are considered; we want the mud area to be considered less costly and easily traversable, but just for a single agent. To do this, follow these steps:

1. Open **Content Drawer** and, in the `Blueprints` folder, create a new Blueprint class of the **Navigation Query Filter** type.

2. Name the asset `NavFilter_MudWalker` and double-click on it to open it.

3. In the **Details** panel, you will see an **Areas** array attribute; click on the + button to add an element.

4. Open the element and do the following:

    *   From the **Area Class** dropdown menu, select **NavArea_Mud**.

    *   Check the **Travel Cost Override** checkbox and set its value to `1.0`.

    *   Check the **Entering Cost Override** checkbox and set its value to `0.0`.

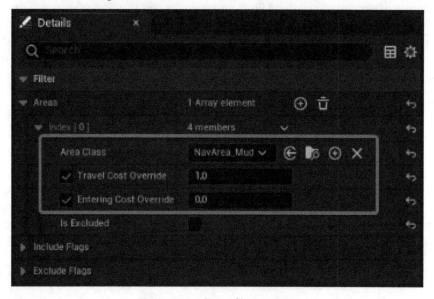

Figure 5.13 – Query filter settings

We are basically creating a different cost for the muddy area of our level – that is, where the **NavArea_ Mud** modifiers are placed.

We now need to modify the AI agent a bit, so it will accept this kind of modifier.

Modifying the agent

We need to make a slight modification to the agent, so open the **BP_NavMeshAgent** Blueprint class and do the following:

1. From the **Move to Actor** node, click and drag on the **Filter Class** incoming pin, and, once released, select **Promote to variable**.

2. Name the newly created variable `FilterClass` and make it **Instance Editable** by checking the corresponding attribute in the **Details** panel.

3. Compile the **Blueprint** and double-check in the **Details** panel that **Default Value** is set to **None**.

Whenever you tell the agent to move to a target point, it will use the `FilterClass` variable – if set to any value – to override the nav mesh cost rules.

Let's test it out in action. Open your gym and select the second agent; then, from the **Details** panel, from the **Filter Class** dropdown, select **NavFilter_MudWalker**.

Test the gym and you will now notice the second agent go straight to the target point, moving through the mud. This cheeky AI agent has gone rogue and decided to play dirty, hasn't it?

In this section, you have discovered the ability to override the cost of a nav mesh, which grants you significant power in creating highly customizable AI characters. With this newfound capability, you can make your AI characters behave distinctively and stand out from the rest of the crowd. I am pretty confident you will understand that this opens up a world of possibilities for creating unique and dynamic gameplay experiences.

In the next section, I will show you another important technique in AI pathfinding, and that's how to make your AI agents avoid each other.

Implementing agent avoidance

It comes as no surprise that, most of the time, you will be working with more than a single agent in a level, and this means there will be a high probability that they will have crossing pathfinding rules; this means that your poor AI entities will be at a significant risk of colliding with each other. As funny as it may be, I guess that's not the intended behavior in your game.

That's why Unreal Engine provides an out-of-the-box – but disabled by default – avoidance system. In this section, we are going to consider how to make AI agents avoid each other.

As always, we will start with a brand-new gym.

Creating the level

As a first step, we will need a gym with some obstacles around. To get started, do the following:

1. From the main menu, select **File | New Level**.

2. Navigate to the Maps/LevelInstances folder and drag an instance of **LI_Lighting** inside your level; set its transform **Location** value to (0, 0, 0).

3. Navigate to the Maps/PackedLevelActors folder and drag an instance of **PLA_Lab_05** inside your level; set its transform **Location** value to (0, 0, 0).

4. Save the level in the Maps folder and name it Gym_NavMesh_06.

This gym is slightly bigger than the previous ones and has some obstacles to make things more interesting. Additionally, there are eight blue tiles – we're going to use eight agents – as depicted in *Figure 5.14*:

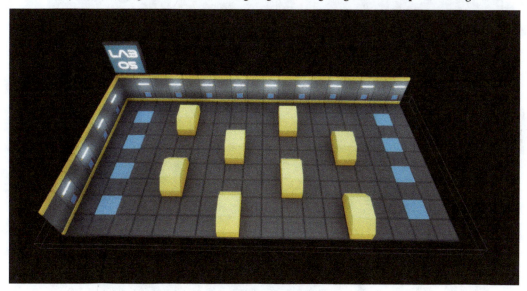

Figure 5.14 – The gym

Adding the agents

As previously mentioned, we are going to add several agents – eight in total – to check how they behave in a crowded environment. So, follow these steps:

1. Add eight instances of the **BP_NavMeshAgent** Blueprint on the level and put each one in a blue tile in the level.

2. Add eight instances of the **NS_Target** Niagara system and put them just behind each blue tile.

3.   For each agent, set the **Target Actor** property value to the **NS_Target** Niagara system that is on the opposite side of the gym. The level should now look like *Figure 5.15*:

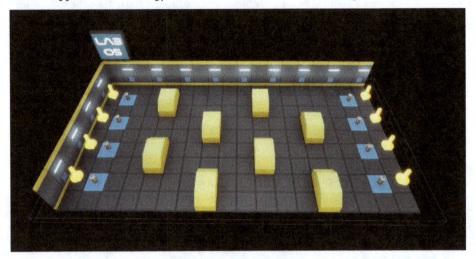

Figure 5.15 – The finished gym

If you try testing the gym, chances are that some of your agents will collide with some other agent moving in the opposite direction, as depicted in *Figure 5.16*:

Figure 5.16 – Ouch, this hurts!

Luckily, our agents are geared up with safety helmets, unlike the characters in your game who might be cruising without any protection! That's why we are going to make the agents' pathfinding slightly cleverer.

Activating avoidance

Once you have opened the **BP_NavMeshAgent** Blueprint, you are ready to enable the avoidance system. To do this, follow these steps:

1. In the **Details** panel, locate the **Character Movement: Avoidance** category and check the **Use RVOAvoidance** property.

2. Set **Avoidance Consideration Radius** to 2000.0.

As stated in *Chapter 3*, *Presenting the Unreal Engine Navigation System*, RVO refers to a feature that enables AI agents to avoid collisions with each other.

When the **Use RVOAvoidance** property is enabled for a character or agent, it allows them to dynamically adjust their movement to avoid colliding with other agents in the environment. The **Avoidance Consideration Radius** property is used to define the radius within which an agent considers other agents for collision avoidance.

Testing the gym at this point will let you see the avoidance system at work; agents will avoid each other while reaching their target point.

Testing a worst-case scenario

Let's test something different; we will be creating a gym where spaces will be a bit more challenging for the agents. The new gym will be a duplicate of the previous one. Start by following these steps:

1. Duplicate the **Gym_NavMesh_06** map and call it Gym_NavMesh_07.

2. Add some obstacles that will create a narrow path, almost forcing the characters to follow a single, specific route. My gym is shown in *Figure 5.17*:

Figure 5.17 – Worst-case scenario

Test the game, and you will witness all the agents diligently making efforts to avoid colliding with one another. You can make a countertest and uncheck the **Use RVOAvoidance** property; you will notice that all the agents initially cluster together and collide with each other before eventually resolving their paths and reaching their respective target points.

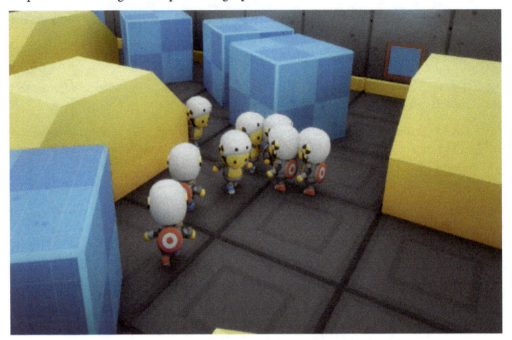

Figure 5.18 – Mass collision

In this section, you have learned how to implement collision avoidance for agents moving on a nav mesh, ensuring that they can successfully navigate while actively avoiding collisions with other agents.

Summary

In this chapter, we were introduced to some more advanced features of the Unreal Engine pathfinding system. Firstly, we saw how to create a nav mesh that can be updated at runtime. Next, we saw how you can override the way an agent interprets the cost of a nav mesh section. Finally, we saw how to use the built-in avoidance system so that AI agents won't collide with each other.

In the upcoming chapter, we will explore the final component of the pathfinding system: debugging and optimization techniques. We will explore ways to identify and resolve any issues that may arise and optimize the system for improved performance. Get ready for an exciting and informative session!

6

Optimizing
the Navigation System

As a project grows in complexity, ensuring that the nav meshes are finely tuned for smooth and efficient AI agent movement becomes crucial. That's why in this chapter, we will delve into the world of optimizing and debugging the Navigation System.

Throughout this chapter, we will explore various techniques and strategies to optimize nav meshes, and we will discuss ways to streamline the pathfinding calculations, reduce computational overhead, and improve overall performance.

You will also acquire the tools and knowledge necessary to effectively debug any issues that may arise within the Navigation System. From troubleshooting pathfinding errors to identifying nav mesh inconsistencies, we will cover a range of debugging techniques to help you conquer any obstacles standing in the way of your agents' smooth navigation.

In this chapter, we will be covering the following topics:

- Understanding the nav mesh debugging tool
- Analyzing nav mesh resolution
- Refining nav mesh generation
- Making further improvements

Technical requirements

To follow the topics presented in this chapter, you should have completed the previous ones and understood their content.

Additionally, if you would prefer to begin with code from the companion repository for this book, you can download the `.zip` project files provided in this book's companion project repository at `https://github.com/PacktPublishing/Artificial-Intelligence-in-Unreal-Engine-5`.

To download the files from the end of the last chapter, click the `Unreal Agility Arena - Chapter 05 - End` link.

Understanding the nav mesh debugging tool

Driven by his relentless pursuit of perfection, Dr. Markus delved into the task of debugging and improving the artificial intelligence dummy puppets. Armed with a tireless spirit and a mind brimming with innovative ideas, he set out to fine-tune the puppets' capabilities and address any glitches or shortcomings they encountered during their experiments.

With his trusty assistant, Professor Viktoria, by his side, Dr. Markus meticulously analyzed the data collected from the puppets' previous expeditions, scrutinizing every line of code, searching for any potential flaws, and seeking opportunities to enhance their performance.

As a game programmer, you already know that optimizing code in your games and hunting down bugs is crucial for creating successful games. The AI system is no exception to this. Luckily, Unreal Engine provides a set of features – the **debugging tools** – that grant you invaluable insights into the inner workings of the Navigation System, allowing you to visualize and analyze the nav mesh in real time. In this section, we will start exploring these tools in order to check how the system behaves and whether it is working well.

> **Note**
>
> In this chapter, we are going to consider just the part of the debugging tools that will let you analyze the nav mesh system, but you should be aware that the AI debugging system covers all the AI features available in Unreal Engine. That's why I will get back to the debugging tools later in this book, when I will be covering other AI topics.

Let's start by checking how the AI debugging tools can be enabled and how to get started with the nav mesh debugging features.

Enabling the AI debugging tools

To enable the debugging tools, all you need to do is to hit the apostrophe key (') on your keyboard. Be aware that on some keyboard layouts – such as mine – this key may not be available; you can add your own shortcut by doing the following:

1. From the main menu, select **Edit | Editor Preferences**.
2. Select the **General – Keyboard Shortcuts** category and, in the search bar, type `show ai debug`.
3. In the **Show AI Debug** field, insert your favorite shortcut. In my case, I opted for the / character from the numpad.

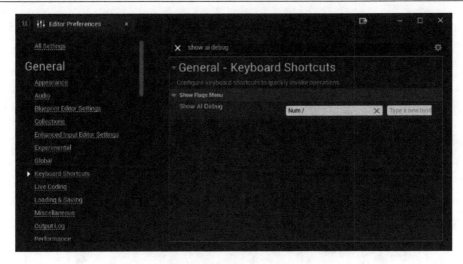

Figure 6.1 – AI debugger shortcut

4. Once the AI debugging tools are enabled, you can toggle the AI information by using the following numpad keys:

- *Numpad 0*: Displays or hides information about the currently available nav mesh data

- *Numpad 1*: Displays or hides general AI debug information

- *Numpad 2*: Displays or hides behavior trees debug information

- *Numpad 3*: Displays or hides EQS debug information

- *Numpad 4*: Displays or hides AI perception debug information

- *Numpad 5*: Displays or hides AI perception system debug information

Let's start checking these tools with the gym levels we have created in the previous chapters.

Inspecting the AI debugging tool

To start using the AI debugging tools, all you need is a level. So, start by opening the **Gym_NavMesh_01** level and doing the following:

1. Disable the nav mesh visualization tool – if enabled – by pressing the *P* key; this will avoid cluttering the AI information display.

2. Start the level simulation and immediately pause it.

3. Enable the AI debugging tools by pressing the apostrophe key; this will open a sidebar and you will get some display messages, as shown in *Figure 6.2*:

File Edit Window Tools Build Select Actor Help
Gym_NavMesh_01
No Active Mode
Gameplay debugger is active.
Clear DebugAI show flag to disable tool
Disable tool
Perspective
Lit
Show
Clear DebugAI show flag to close, use Numpad to toggle categories.
0:Navmesh 1:AI 2:BehaviorTree 3:EQS 4:Perception 5:PerceptionSystem

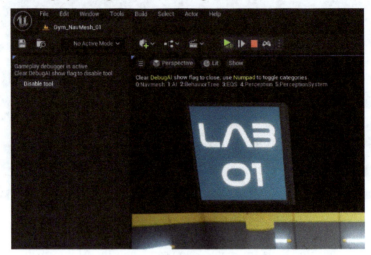

Figure 6.2 – AI debugging tools in action

4. Once the debugging tools are enabled, ensure that the **Navmesh** and **AI** categories are enabled and that all the other ones are disabled by using the corresponding numpad keys. Enabled categories are highlighted in green, as shown in *Figure 6.3*:

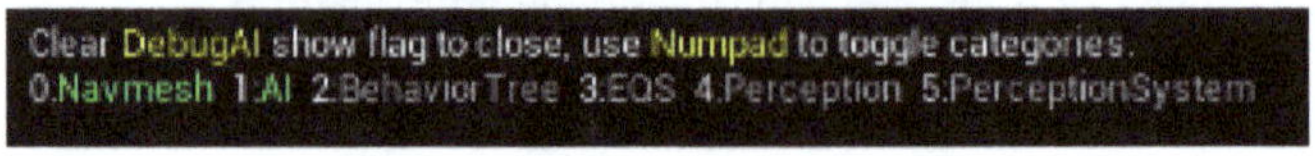
Clear DebugAI show flag to close, use Numpad to toggle categories.
0:Navmesh 1:AI 2:BehaviorTree 3:EQS 4:Perception 5:PerceptionSystem

Figure 6.3 – Enabled categories

Still with the game paused, do the following:

5. Look for **BP_NavMeshAgent**, and you will notice that it now has a red icon associated.
6. Click on the actor to select it; this will show some information on it. The most important one is the associated AI controller.

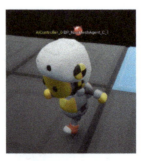

Figure 6.4 – AI agent

Additionally, on the display, you will see additional information about the AI agent, as shown in *Figure 6.5*:

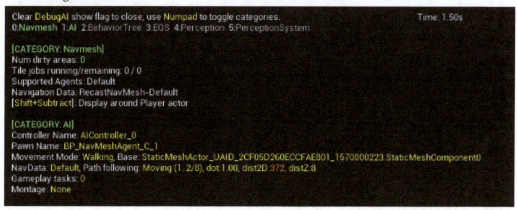

Figure 6.5 – AI agent information

As you can see from the image, this display offers a plethora of information, including the controlling pawn, the controller, and the mesh the agent is walking on.

Finally, you will get probably one of the most interesting pieces of information on the nav mesh and the pathfinding system; if you move around the level, you will see a highlight of the polygons that make up the path to the target point.

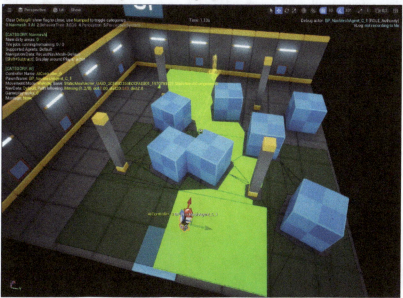

Figure 6.6 – Pathfinding polygons

As an extra exercise, you can open up all the other levels and see which information you can get from the tool. For example, you may look at the **Gym_NavMesh_04** level and check what happens when the moving platform moves up and down.

In this section, you have been introduced to the tools that will provide meaningful pieces of information on the AI system; in the next section, I'll show you how to derive insights on nav mesh generation for your levels in order to optimize them.

Analyzing nav mesh resolution

Unreal Engine provides a **Navigation Mesh Resolution** system that lets developers create mesh tiles at three distinct levels of detail within a single nav mesh. This means that you have the flexibility to generate sets of tiles with high, medium – the default option – or low precision settings. By opting for different precision levels, users can achieve faster generation – in terms of computing time – of a dynamic nav mesh while the game is playing.

> **Note**
>
> When we speak about **resolution**, we mean the precision and quantity of cells produced to map a specific navigation area.

A high-resolution tile may divide a given area into more polygons to closely approximate its shape. Conversely, a low-resolution tile will encompass the same area but with fewer polygons. This trade-off enables quicker tile generation but may sacrifice some accuracy in the process.

As a first test, we can start analyzing one of our gyms:

1. Open the **Gym_NavMesh_01** level.

2. For a better view of the level, select the **Top** view.

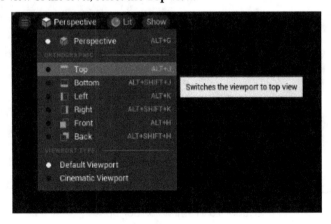

Figure 6.7 – Selecting the Top view

3. In **Outliner**, select the **Recast Nav Mesh** actor.

    To get the overall information on the cost for your nav mesh, do the following:

4. With the **Recast Nav Mesh** actor still selected, check the **Draw Tile Bounds** property in the **Details** panel.

5. Check the **Draw Tile Bounds Times** property. You should get a view like that shown in *Figure 6.8*:

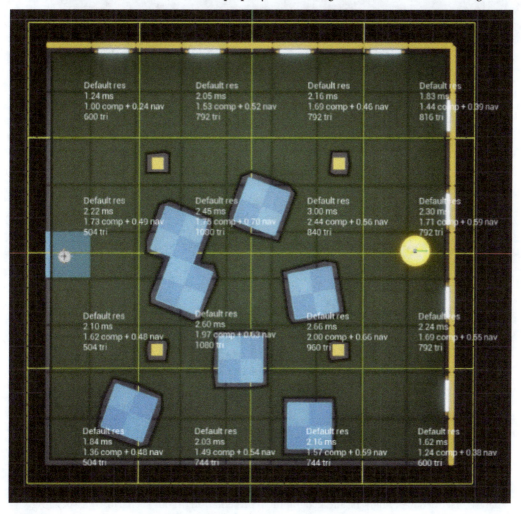

Figure 6.8 – Tiles and tile generation time

The **Draw Tile Bounds Time** property shows the time cost of processing a specific tile; you will notice that more complex areas will take more time to compute. If you want a more visual representation of the cost, you can check **Draw Tile Build Times Heat Map**, which will enable a heat map visualization of the cost, as shown in *Figure 6.9*:

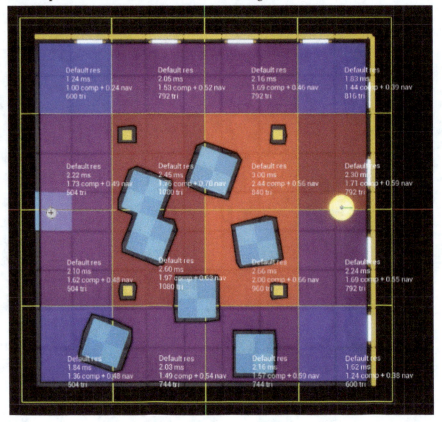

Figure 6.9 – Heat map visualization

Note

If you are not familiar with this feature, a heat map is a visual representation of data that employs color-coded systems to display the volume or intensity of a specific attribute across a defined area or dataset. Heat maps are adaptable and find application in diverse scenarios, including data analysis, user behavior analysis, and geographic representation. In our scenario, by associating colors with different levels of intensity – with blue being the less costly area and lighter colors being the higher costly ones – heat maps illuminate regions of increased or decreased traversal cost, enabling observers to discern patterns and crucial circumstances.

This view is particularly useful when dealing with large areas and wanting to obtain an overview of the more resource-intensive zones. In the preceding figure, the red zones are the ones that will cost more in navigation terms.

In this section, you have grasped the process of analyzing a nav mesh to identify potential issues; in the next section, we will explore strategies for addressing and enhancing its generation.

Refining nav mesh generation

It should come as no surprise that the complexity of an environment directly affects the time it takes for the system to generate a nav mesh. Conversely, if the generation time is shorter, the resulting nav mesh may be less precise. In most cases, your final system will involve a trade-off between computation speed and precision. This means it is mandatory, as a game developer, to understand how to properly set up your nav mesh generation.

In this section, I will provide you with some advice on how to optimize the way a nav mesh is generated.

Influencing nav mesh resolution

The next test we will perform involves modifying the resolution of a nav mesh. Start by following these steps:

1. Open the **Gym_NavMesh_02** level.
2. If it is not already enabled, select the **Top** view.
3. In the **Outliner** panel, select the **Recast Nav Mesh** actor.
4. In the **Details** panel, check the following properties:

    - **Draw Tile Bounds**
    - **Draw Tile Bounds Times**
    - **Draw Tile Build Times Heat Map**

You should get a view similar to *Figure 6.10*:

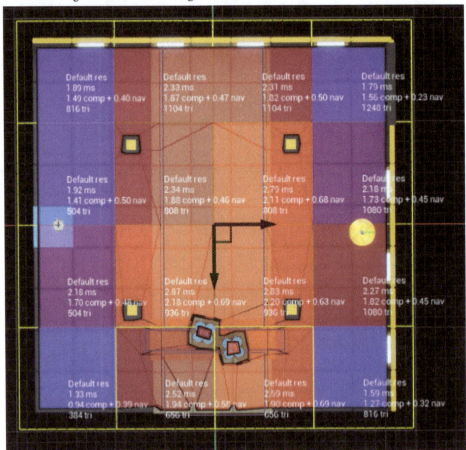

Figure 6.10 – Heat map visualization

As you can see from the heat map, the central part has a higher cost. However, the overall geometry is quite simple; we have the muddy area and a bridge, which are basically rectangular, non-rotated areas. Let's lower the cost for our mesh. Start doing the following steps:

5. In the **Outliner** panel, select the three **Nav Modifier Volumes** – the two muddy areas and the bridge area.

6. In the **Details** panel, search for the **Nav Mesh Resolution** property and, in the dropdown menu, select **Low**.

You will notice that the time cost for the central part of the map will immediately drop, as depicted in *Figure 6.11*:

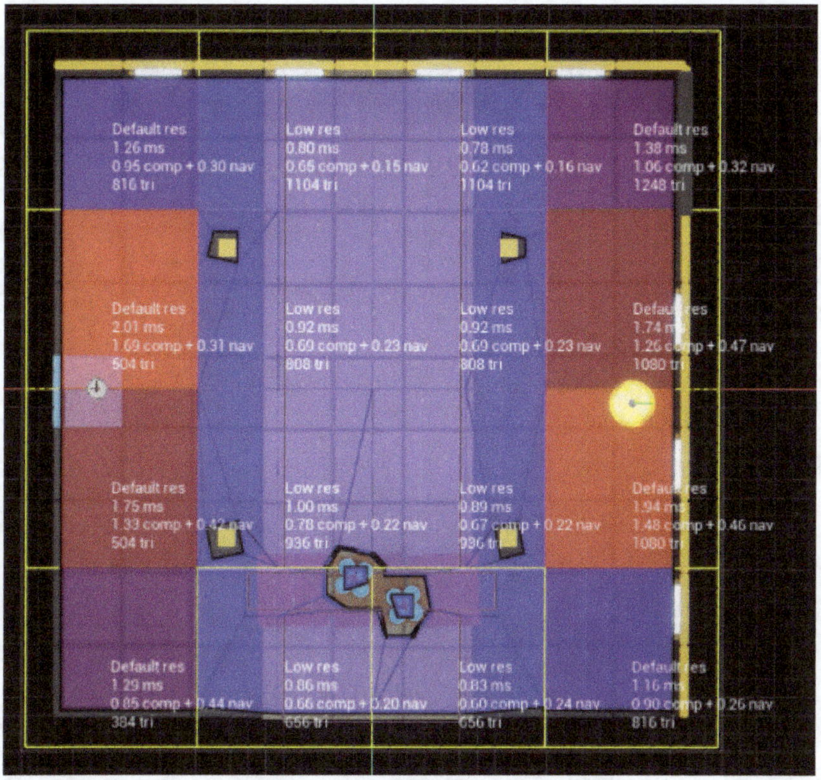

Figure 6.11 – Improved nav mesh

When you look at the heatmap visualization, you might see some tiles that have been left untouched by the previous modification – the ones on the sides – changing color and appearing more expensive. However, don't let this trick you – the system will simply highlight areas that will be more costly in terms of the overall nav mesh. In fact, if you check those areas, you will see that the cost hasn't changed at all; they have simply become more costly than their central map counterparts. This means that the cost for a particular area is relative to all other areas analyzed by the system.

As another experiment, you can set the bridge **Nav Mesh Modifier** actor to a **Nav Mesh Resolution** value of **High** and see the results. Spoiler alert! The tiles, including the bridge, will become the most expensive parts of the level!

Changing nav mesh resolution

If, at this point, you are wondering whether you can adjust the nav mesh resolution for your level, the answer is an unsurprising yes!

By selecting the **Recast Nav Mesh** actor, in the **Detail** panel, you will find a **Nav Mesh Resolution Params** option with three fields – **Low**, **Default**, and **High** – that will let you decide the cell size for your level.

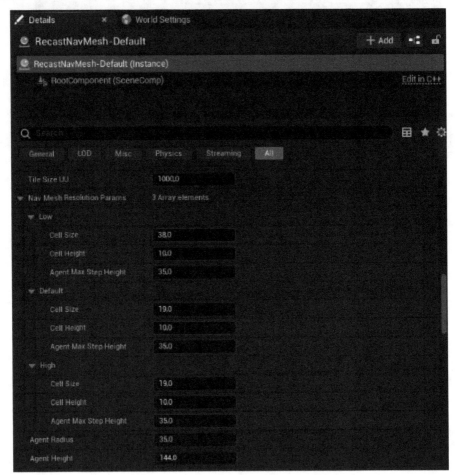

Figure 6.12 – Nav Mesh Resolution Params

As a rule of thumb, just remember that the higher the cell size, the less costly the zone will be.

Changing the tile size

You are already aware that the nav mesh is partitioned into tiles, which are utilized to reconstruct specific sections of the nav mesh itself. As each tile consists of cells, rebuilding tiles entails recreating all its cells with the updated information.

Larger tiles encompass more cells, resulting in a higher cost to rebuild than smaller tiles. However, when processing a tile, the system also handles the adjacent cells along the tile's edges. It is important to consider this additional overhead cost when determining the size of your tiles. In certain scenarios, the cumulative overhead cost of processing numerous smaller tiles may exceed the cost of rebuilding a single large tile. Therefore, careful consideration should be given when choosing the appropriate tile size to optimize performance.

To achieve optimal performance during tile rebuilding at runtime, Epic Games' recommendation is that you set each **Cell Size** property – **Low**, **Default**, and **High** – as a multiple of each other and set the **Tile size UU** property to a value that is divisible by all **Cell Size** values. For instance, in *Figure 6.13*, I have set the values for the **Gym_NavMesh_02** level as follows:

- **Tile Size UU** as `960.0`
- **Low Cell Size** as `60.0`
- **Default Cell Size** as `30.0`
- **High Cell Size** as `15.0`

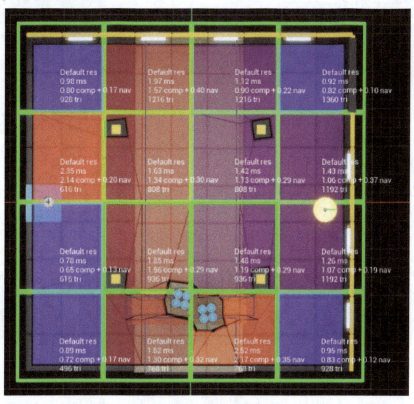

Figure 6.13 – Cell size example

As you can see, the **Low Cell Size** value can be divided by **Default Cell Size** and **High Cell Size**, giving an integer value. The same can be said for **Tile Size UU**, which can be divided by **Low Cell Size**, **Default Cell Size**, and **High Cell Size**.

As an extra example, *Figure 6.14* shows the same level with these values:

- **Tile Size UU** as 1280.0

- **Low Cell Size** as 80.0

- **Default Cell Size** as 40.0

- **High Cell Size** as 20.0

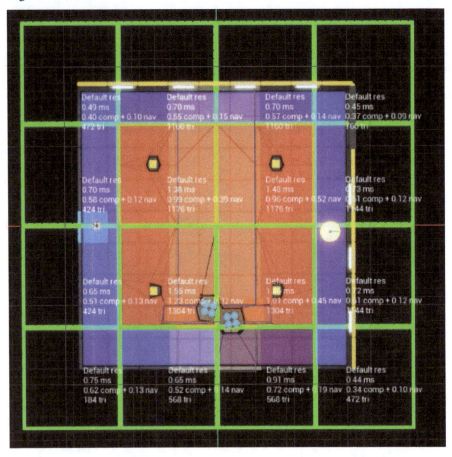

Figure 6.14 – Cell size alternative example

In the second example, a larger tile size is employed. However, it is evident that the optimization is not as effective, leading to inefficient utilization of space.

In this section, I have presented you with some valuable tips on how to analyze your nav mesh generation and effectively optimize it by changing cell resolution and tile size. In the next section, I will give you some additional advice on how to make your navigation even better.

Making further improvements

Tweaking a nav mesh may take a long time – depending on what you want to achieve –and sometimes, it's a matter of trial and error and personal experience. In this section, I will give you additional advice on how to improve your maps to make them more functional.

Tweaking resolution

Choosing the right nav mesh resolution is not just a matter of computational performance; sometimes it may even affect your agent navigation.

As an example, take into consideration *Figure 6.15* showing a portion of the **Gym_NavMesh_07** level:

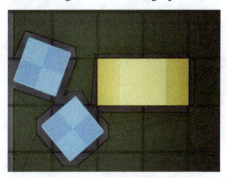

Figure 6.15 – Nav mesh resolution

In this case, the **Default Cell Size** value has been set to 20.0 and, as you can see, there's no walkable area between the obstacles. However, if you decrease the value to 5.0, you will get a totally different scenario, as depicted in *Figure 6.16*:

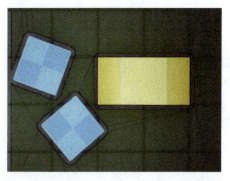

Figure 6.16 – Improved nav mesh resolution

As you can clearly see, there are now open passages between the obstacles; the higher the coin, the finer the wine!

Disabling mesh influence

At times, your nav mesh may become cluttered with meshes that could lead to unreachable pathways, yet they are included in the nav mesh generation process. In such instances, it is advisable to hide them to prevent them from affecting generation time. As an example, consider the situation depicted in *Figure 6.17*:

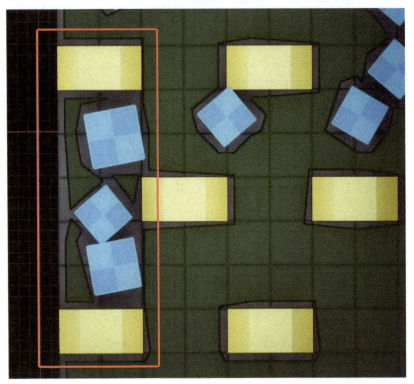

Figure 6.17 – Unreachable path

As you can see, the area in the red rectangle is unreachable, yet the central cube will be counted during nav mesh generation. This means that, even though it won't influence any pathfinding resolution, it will have to be computed anyway. In this case, you can exclude it by completing the following steps:

1. Select the mesh.

2. In the **Details** panel, look for the **Can Ever Affect Navigation** property and uncheck the checkbox.

    The mesh will now be excluded from the nav mesh generation, but the overall result will remain the same, as you can see from *Figure 6.18*:

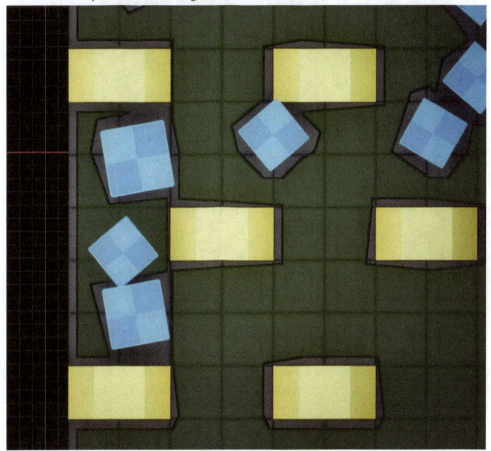

Figure 6.18 – The improved area

You can go even further by doing the following:

3. Select the other two boxes and uncheck the **Can Ever Affect Navigation** property for both.

4. Add a Nav Modifier Volume actor and scale it so that it will encompass the whole area.

5. Set the volume **Area Class** property to **NavArea_Null**.

Figure 6.19 shows the final result:

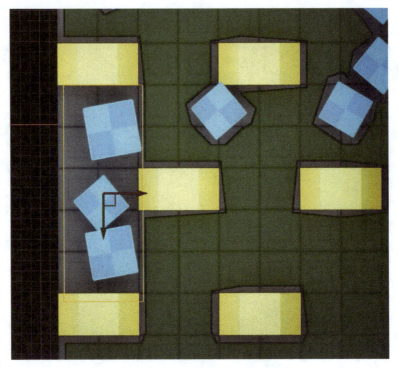

Figure 6.19 – The optimized area

As you can see, we are now using a single modifier to avoid the influence of three meshes on the final nav mesh generation. Picture all the time you'll have spared once your level is jam-packed with props and obstacles of every kind!

Summary

In this chapter, we went through some more advanced techniques to help improve your nav mesh generation. Starting with the debugging tool, we then learned how to analyze and tweak the mesh generation process. Finally, we learned a few tricks that will help us craft more effective levels.

With this, *Part 2* of this book comes to an end; starting from the next chapter, get ready to face probably one of the most interesting topics when speaking about AI in game development: behavior trees.

Be prepared. Your AI buddies are about to get a major overhaul and will never be the same!

Part 3: Working with Decision Making

In the third part of this book, you will get a comprehensive introduction to the powerful and versatile behavior tree system within the Unreal Engine framework. Additionally, you will be presented with advanced features so that you can implement your own complex game AI logic.

This part includes the following chapters:

- *Chapter 7, Introducing Behavior Trees*
- *Chapter 8, Setting Up a Behavior Tree*
- *Chapter 9, Extending Behavior Trees*
- *Chapter 10, Improving Agents with the Perception System*
- *Chapter 11, Understanding the Environment Query System*

Introducing Behavior Trees

In the universe of game development, **behavior trees** are hierarchical structures that govern the decision-making processes of AI characters, determining their actions and reactions during gameplay. As a game programmer, delving into the intricacies of behavior trees is crucial, as it will empower you with the ability to craft dynamic, intelligent, and engaging virtual entities that enhance the player's game experience.

This chapter aims to provide a gentle introduction to behavior trees and Blackboards, as well as their application within the Unreal Engine.

In this chapter, we will be covering the following topics

- Explaining behavior trees
- Understanding behavior trees in Unreal Engine
- Understanding the Blackboard

Technical requirements

There are no technical requirements to follow this chapter.

Explaining behavior trees

In its broader sense, a behavior tree is a mathematical model used in many fields of computer science, including video games. It outlines the transition between a finite set of tasks in a modular manner. The power of behavior trees lies in their ability to create intricate tasks from simple components, without going into the details of how each component is implemented. While behavior trees share some similarities with hierarchical state machines – where states are organized in a hierarchy, allowing for better reuse of behaviors – the primary distinction lies in the fact that tasks, not states, serve as the fundamental elements of behavior. The main advantage is their intuitive nature, making them less prone to errors; this is why they are highly favored within the game development industry.

Today's video games are increasingly intricate, leading to a proportional complexity in AI characters. Hence, the maintenance of these characters – or agents – is crucial. Unlike systems such as finite state machines, which become difficult to maintain as the number of states increases, behavior trees offer a practical and scalable solution for decision-making processes. When an agent executes a behavior tree, it conducts a **depth-first search** to locate and execute the lowest-level leaf node.

The key advantages of behavior trees over other systems lie in their scalability, expressiveness, and extensibility. Unlike other systems, behavior trees do not involve explicit transitions between states; instead, each node in the tree specifies how to run its children. This stateless nature eliminates the need to track previously executed nodes to determine the next set of behaviors. The expressiveness of behavior trees stems from the use of various levels of abstraction, implicit transitions, and complex control structures for composite nodes.

Furthermore, in behavior trees, transitions occur through calls and return values exchanged between tree nodes, facilitating a two-way control transfer mechanism.

Behavior tree structure

A behavior tree is visually depicted as a tree structure with nodes categorized as **root, control flow**, and **execution** – or **tasks**. In this representation, each node may have a parent node and one or more children. In particular, the following is worth noting:

- The root node has no parents and only one child
- Control flow nodes have one parent and at least one child
- Execution nodes have one parent and no children

A behavior tree is executed starting from the root, which sends execution triggers to its child nodes.

Whenever a control flow node is reached, it will control the execution and flow of decision-making within the tree, determining which tasks or sub-trees should be executed based on certain conditions or rules.

Every time an execution node is triggered, it will execute a specific task, reporting back to its parent with a status of *running* if the task is ongoing, *success* if the objective is accomplished, or *failure* if the task is unsuccessful.

Figure 7.1 shows an example of a behavior tree execution, starting from the root, going to a control flow node, and finally executing a task:

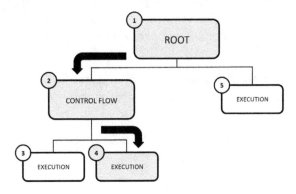

Figure 7.1 – Behavior tree example

> **Note**
>
> Behavior tree nodes are executed from top to bottom and left to right; this is also the way they are usually numbered.

It goes without saying that there isn't just one way to implement behavior trees. That's why in the next section, I'll dive into all the nitty-gritty details about the Unreal Engine system.

What is a behavior tree in Unreal Engine?

In Unreal Engine, behavior trees are assets that are edited in a similar way to Blueprints – that is, visually – by adding and linking a set of nodes with specific functionalities to form a **behavior tree graph**. During the execution of logic in a behavior tree, a separate asset known as a Blackboard – further details on this will be provided later in this chapter – is used to retain information that the behavior tree requires to make well-informed decisions.

A behavior tree is handled by a `BehaviorTreeComponent` instance that is held by the `AIController` instance. It should be noted that the component is not automatically attached to the controller; you will need to add it through C++ or Blueprints. If no component is present, it will automatically be created at runtime.

When comparing Unreal Engine behavior trees with other behavior tree systems, one key distinction to keep in mind is their event-driven nature, which prevents constant code execution. Instead of continuously checking for relevant changes, an Unreal Engine behavior tree listens for events that can prompt tree modifications. Using an event-driven architecture provides performance enhancements and debugging capabilities benefits – this is something I will show in the upcoming chapters.

Behavior tree node instancing

It needs to be noted that behavior trees exist as **shared objects** in your project; this means that all agents using a behavior tree will share the same instance, and all shared objects will be unable to store agent-specific data. The main advantages of using shared nodes are CPU speed improvement and reduced memory usage.

Agent-specific data can be leveraged in many ways – one being the Blackboard that we will see later in this chapter – to give you more flexibility on how to use your behavior tree.

Another such method is instancing single nodes; this will grant each AI agent using a behavior tree a unique instance of the node at the cost of higher performance and memory usage. An example of a node using such a method is the **PlayAnimation** task.

Order of execution

As previously mentioned, behavior tree nodes are executed from top to bottom and left to right and Unreal Engine is no exception. Nodes are numbered following this convention to easily track the execution order. *Figure 7.2* shows a behavior tree from the **Lyra Starter Game** project, showing the nodes with their corresponding sequence numbers:

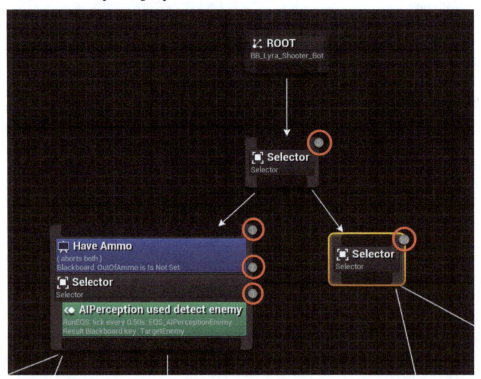

Figure 7.2 – Behavior tree sequence

> **Note**
> In Unreal Engine, a root node is never numbered because it is not considered part of the sequence.

In this section, you have received a gentle introduction to behavior trees and how they are executed. The next section will delve deeper into the Unreal Engine system to help you better understand how to incorporate behavior trees effectively into your games.

Understanding behavior trees in Unreal Engine

Understanding behavior trees and what they are made of is essential for designing effective AI systems in Unreal Engine; in this section, I will be presenting you with the key concepts associated with behavior trees to help you start developing your own AI characters.

In Unreal Engine, there are five types of elements in behavior trees:

- Root node
- Task nodes
- Composite nodes
- Decorators
- Services

To provide you with a comprehensive understanding of each type, I will present them individually, ensuring a clear depiction of their respective functions.

The root node

The root node functions as the initial point for a behavior tree; it holds a distinct position within the tree and is governed by a set of special rules:

- There can be only one such node in the tree structure
- It can have only one connection, and if this connection is removed, then the entire tree is disabled
- It does not support the attachment of decorator or service nodes

Figure 7.3 – Root node

Task nodes

Task nodes are responsible for performing actions such as moving an AI or adjusting Blackboard values. A task will not stop its execution until a failure or success result is reported.

A task node can also have one or more decorators or services attached, allowing for more complex behaviors and interactions within the game environment.

> **Task nodes**
> Tasks are identified by a purple color.

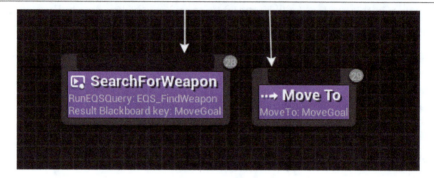

Figure 7.4 – Task examples

Unreal Engine includes a set of pre-built tasks that are readily available for use. These tasks address the most generic scenarios that developers are likely to require. However, tasks can be extended to allow you to create your own custom nodes.

Here is a partial list of some of the tasks that will be available to you as standard features:

- **Finish With Result**: Once executed, this node will instantly finish with a defined result – `Succeeded`, `Failed`, `Aborted`, or `In Progress`

- **Move To**: Once executed, it will move the AI agent to a target location by using the Navigation System

- **Move Directly Toward**: Once executed, it will move the AI agent to a target location without using the Navigation System

- **Wait**: Once executed, it will cause the behavior tree to wait on this node until a specified time has passed

- **Play Animation**: Once executed, this node will play a specified animation asset

- **Play Sound**: Once executed, this node will play a specified sound

As you can see, tasks represent individual actions or operations that an AI agent can perform; you can use them to create simple actions or combine several to create more complex behaviors.

Composite nodes

Composite nodes define the root of a branch and set the rules for its execution; additionally, they are the only nodes that can be applied to the root node of a behavior tree.

A composite node can also have decorators and services applied, enabling more complex logic in it. Once a service is applied, it will be active while the children of the composite are executed.

> **Composite nodes**
> Composites are identified by a grey color.

There are three composite nodes available:

- **Selectors**
- **Simple parallels**
- **Sequences**

Let's examine them one by one.

Selectors

Selector nodes execute their children sequentially from left to right, and they will halt execution as soon as one of them succeeds. When a child of a selector node succeeds, the selector itself is considered successful. On the other hand, if all the selector's children fail, the selector node itself is marked as failed.

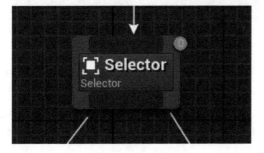

Figure 7.5 – Selector node

Simple parallels

Simple parallel nodes enable the execution of a single main task node concurrently with a complete tree. After the main task is completed, you may decide – through the **Finish Mode** attribute – whether the node should immediately finish, halting the secondary tree, or whether it should wait for the secondary tree to finish before completing.

Figure 7.6 – Simple Parallel node

Sequences

Sequence nodes run their children sequentially from left to right. They halt execution when a child fails. If a child fails, the sequence also fails. Unlike selectors, the success of the sequence is achieved only when all its children succeed.

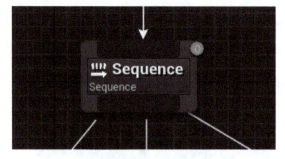

Figure 7.7 – Sequence node

Decorators

Decorators – sometimes also referred to as conditionals – are used to determine whether a branch in the tree, or even a single node, can be executed. They must be attached to either a composite or a task node.

Decorators play a crucial role in determining the execution path of branches within the behavior tree; they essentially act as decision-makers, evaluating whether a specific branch or individual node should proceed with execution. They serve as a condition, assessing the viability of continuing down a particular branch, signaling a precautionary failure if the task – or a sub-tree – is destined to fail. This preemptive action helps prevent the decorator from attempting to carry out a task – or sub-tree – that is bound to fail due to various reasons such as insufficient information or outdated objectives.

> **Decorators**
>
> Decorators are identified by a blue color.

Figure 7.8 – Decorator applied to a Selector

Unreal Engine includes a set of pre-built decorators that are readily available for use, but they can be extended to allow you to create your own custom ones.

Here is a partial list of some of the tasks that will be available to you as standard features:

- **Blackboard**: Will check whether a value has been set – or not set – on a given Blackboard key
- **Composite**: Will let you create some more advanced logic than built-in nodes by using **AND**, **OR**, and **NOT** nodes
- **Cooldown**: Will lock the execution of a node or a branch until a predefined time has passed
- **Does Path Exists**: Will check whether a path is found between two points
- **Loop**: Will loop a node or a branch indefinitely – or a number of times, if set

Most decorators include an **Inverse Condition** property, that will let you, well... invert the condition, giving you more flexibility. As an example, you may use the same decorator in a behavior tree to execute different tasks under opposite conditions.

For instance, you may use **Does Path Exist** to move an AI agent to a target point and use **Inverse Condition** on another instance of **Does Path Exist** to look for an alternative target point.

In conclusion, decorators serve as decision points that determine whether a certain action or branch within the behavior tree should be executed or not.

Services

Services can be attached to composite or task nodes and run at specific intervals – defined in the **Interval** attribute – while their branch is active. They are commonly employed for conducting checks and updating the Blackboard.

Once triggered by a task or a composite, a service will keep executing regardless of the number of parent-child levels being executed below the owing node.

> **Services**
>
> Services are identified by a green color.

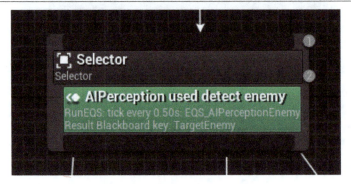

Figure 7.9 – Service applied to a Selector

Services are very specific to the behavior tree you are developing; this means that, most probably, you will have to create your own custom ones. However, Unreal Engine provides two pre-built services that are readily available for use:

- **Default Focus**: This enables quick access to an actor from the AI controller instead of using a Blackboard key.

- **Run EQS**: This can be used to regularly execute an EQS – more on this in *Chapter 11, Understanding the Environment Query System* – at assigned intervals. It can also update a specified Blackboard key.

Having covered the various types of nodes that make up a behavior tree, it's now time to get into the next section in order to delve into Blackboard assets.

Understanding the Blackboard

In Unreal Engine, the Blackboard is a crucial component of behavior trees; it acts as a memory space – some sort of brain – where AI agents can read and write data during their decision-making process. This means that developers will be able to query and update information stored within it.

The Blackboard is created as a **Blackboard Data** asset, which will be assigned to a behavior tree, and it contains a set of variables – named keys – that store specific information of predefined types. These keys can be accessed and manipulated during runtime to influence the decision-making of AI characters.

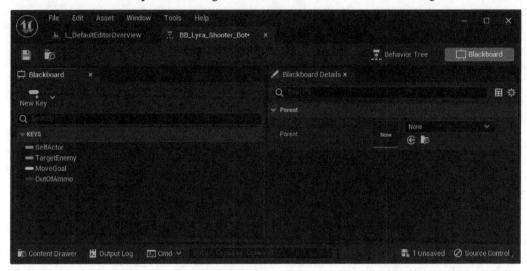

Figure 7.10 – Blackboard example

A key can be set to **Instance Synced**; in this case, the key itself will be synchronized across all instances of the Blackboard. This synchronization ensures that any changes made to the value of the key will be reflected consistently across all instances of the AI agents sharing the same behavior tree and Blackboard.

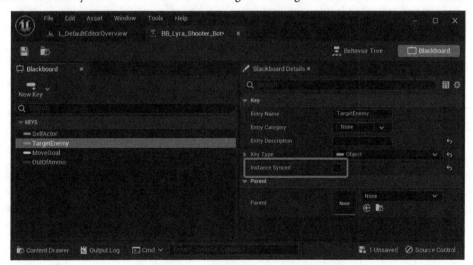

Figure 7.11 – The Instance Synced property

A Blackboard can store up to 255 keys and it supports the following data types:

- `FVector`
- `FRotator`
- `Bool`
- `Int32`
- `Float`
- `UClass`
- `UObject`
- `FName`
- `UEnum`
- `FString`

> **Note**
> A Blackboard cannot store arrays.

The `BlackboardComponent` instance will allow you to query data from a Blackboard and store data in the Blackboard itself. The creation system follows a similar pattern to the `BehaviorTreeComponent`, as explained previously in this chapter.

Despite its apparent simplicity, understanding how a Blackboard works is crucial for ensuring the effective functioning of your AI agents.

Summary

In this chapter, we were introduced to the behavior tree system. Beginning with a brief theoretical overview, we saw how behavior trees function in Unreal Engine and learned about the key components that constitute the entire system. Finally, we discussed the Blackboard asset, a crucial element for the effective operation of behavior trees.

Get ready for the upcoming chapter, where we will guide you back into action and make you craft a behavior tree for our very own dummy character. Specifically, you will be creating your own custom services and tasks in order to give our soon-to-be AI agent a proper AI brain. Brace yourself, because things are about to get intriguingly wild and – sometimes – delightfully chaotic!

8

Setting Up a Behavior Tree

Understanding how to properly set up a behavior tree is crucial for developing effective AI systems in your games. As you have seen from the previous chapter, behavior trees serve as a powerful tool to define the logic and decision-making processes of AI characters, allowing developers to create complex behaviors in a structured and manageable manner. This chapter will provide valuable insights into the fundamentals of behavior trees' implementation and their inner workings in Unreal Engine.

In this chapter, we will be covering the following topics:

- Extending the Unreal Agility Arena
- Creating a behavior tree
- Implementing behavior tree tasks and services
- Setting up a behavior tree on an agent

Technical requirements

To follow along with this chapter, you'll need to use the starter content available in this book's companion repository, located at `https://github.com/PacktPublishing/Artificial-Intelligence-in-Unreal-Engine-5`. Through this link, locate the section for this chapter and download the `Unreal Agility Arena - Starter Content` ZIP file.

If you somehow get lost during the progress of this chapter, in the repository, you will also find the up-to-date project files at `Unreal Agility Arena - Chapter 08 End`.

Also, to fully understand this chapter, it is necessary to have some basic knowledge about Blueprint visual scripting and C++; as an extra piece of advice, you may want to take a peek at *Appendix A, Understanding C++ in Unreal Engine*, for a gentle introduction (or a refresher) on the C++ syntax in Unreal Engine.

Extending the Unreal Agility Arena

To get started, let's continue exploring the short novel we introduced in *Chapter 4, Setting Up a Navigation Mesh*:

As Dr. Markus and his trusty assistant Professor Viktoria continued to refine their AI dummy puppets, they stumbled upon an intriguing challenge: the limited power supply of the puppets' batteries. It seemed that the advanced AI technology consumed energy at an alarming rate, causing the puppets to shut down unexpectedly.

Undeterred by this setback, Dr. Markus saw an opportunity to turn this limitation into a unique aspect of the puppets' behavior. He theorized that the puppets' interactions, when powered by dwindling battery life, would mimic human fatigue and exhaustion. With great excitement, Dr. Markus and Professor Viktoria devised a plan to create a new series of experiments centered around the puppets' limited power supply.

After learning all that information in the last chapter, it's time to dive in and begin crafting your own AI agents, equipped with fully functional behavior trees. To keep things simple and clean, I will start with a brand-new project, but you are free to continue developing the work you started in the previous chapters.

To start, I will give you some short information on what will be created.

Updating the project brief

As a starting point, you will need to create a new dummy character (the one from previous chapters is too limited) that will need to implement some base logic. This will let you extend its basic functionality once we start creating more advanced characters.

The main requisites are listed as follows:

- The AI agent will be implemented in C++
- It will have the ability to move at two different speeds – walking and running
- It will be provided with a battery system that will consume energy when walking and recharge when standing still
- It will need to be controlled by a custom `AIController` class that will use behavior trees

Once the character has been created, we will be able to start creating new gym levels to create and test new AI agent behaviors. So, let's start by creating the project.

Creating the project

The project creation process is basically the same as the one we covered in *Chapter 4, Setting Up a Navigation Mesh*, so I won't go much into detail about it; the distinction we will make (which is certainly not trivial) is the inclusion of C++ classes.

Luckily, this is going to be a seamless transition, as when you create a new C++ class for the first time, Unreal Engine sets up the whole system for you. Once the C++ project files have been generated, you should see the C++ Classes folder in your **Content Browser** window, along with the Content folder.

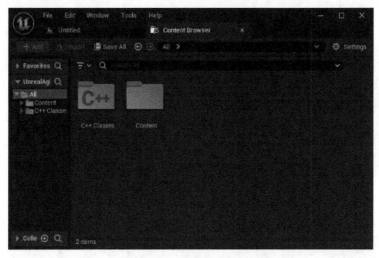

Figure 8.1 – The C++ folder

> **Note**
>
> If the C++ Classes folder does not appear in your **Content Browser** window, you will probably need to open the **Settings** window and tick the **Show C++ Classes** option, as shown in *Figure 8.2*.
>
>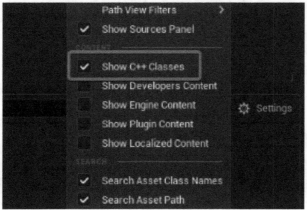
>
> Figure 8.2 – Enabling the C++ folder

Let's start creating the character class.

Creating the character

The first thing we are going to do is to create the base character class for our prospective AI dummies. To do so, follow these steps:

1. From the main menu of the Unreal Engine Editor, select **Tools | New C++ Class**.

Figure 8.3 – Creating a C++ class

2. From the **Add C++ Class** pop-up window, select the **Character** option and click **Next**.

Figure 8.4 – Class selection

3. In the following window, insert `BaseDummyCharacter` into the **Name** field and leave the rest as it is.

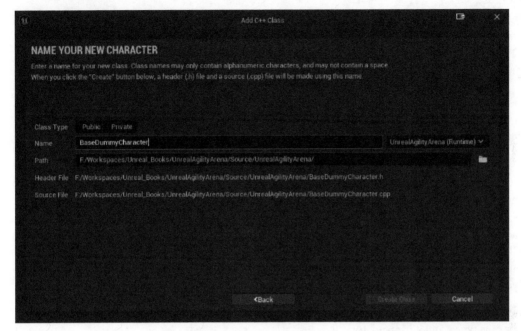

Figure 8.5 – Class creation

4. Click the **Create Class** button to start the class creation process.

As this is the very first class you have created, Unreal will start generating the C++ project; after that, your IDE – most probably Visual Studio or Rider – will open, and you will be able to start developing your class.

Handling the battery status

Before implementing the character, we need to define how its battery status will be tracked; that's why we are going to create a simple enumeration class that will list how much the battery is charged. To do this, follow these steps:

1. Inside your IDE, locate the `UnrealAgilityArena/Source/UnrealAgilityArena` folder.

2. Inside this folder, create a new text file and call it `BatteryStatus.h`.

3. Open the file to edit it.

Inside the file, add this piece of code:

```
#pragma once

UENUM(BlueprintType)
enum class EBatteryStatus : uint8
{
    EBS_Empty = 0 UMETA(DisplayName = "Empty"),
    EBS_Low = 1 UMETA(DisplayName = "Low"),
    EBS_Medium = 2 UMETA(DisplayName = "Medium"),
    EBS_Full = 3 UMETA(DisplayName = "Full")
};
```

I guess you are already familiar with what an enumeration does, but a short explanation here is mandatory; we are defining an enum class, which in Unreal Engine needs to be of type uint8, and we are listing four levels of battery charging – Empty, Low, Medium and Full. The UENUM() macro defines an enum class for the Unreal Engine framework, and the BlueprintType specifier will expose it to the Blueprint system, making it usable for variables. The DisplayName metadata defines how the value will be shown in the Blueprint system; you can use your own convention here if you so desire.

Finally, pay attention to the *E* prefix in the name definition; this is the Unreal Engine naming convention for enum types, and it is mandatory.

With the battery status defined, we are ready to start implementing the dummy character.

Implementing the character

To start implementing the AI agent, open the BaseDummyCharacter.h file – that is, the header for the agent class – and start adding the class definitions.

As a first step, add the battery status definition at the top of the file, where all the #include definitions are set:

```
#include "BatteryStatus.h"
```

> **Note**
>
> All the #include declarations you'll be adding always need to be put before the .generated.h definition – in this case, BaseDummyCharacter.generated.h. The .generated.h definition always needs to be the last in the declaration list; this convention ensures that all the necessary dependencies are properly set at compilation time.

Declaring properties

The second step is to add all the properties that will be exposed in the extending Blueprint classes. To do so, under the `public` section of the file, add the following declarations:

```
UPROPERTY(EditAnywhere, BlueprintReadWrite, Category = "Dummy
Character")
float MaxBatteryLevel = 100.f;

UPROPERTY(EditAnywhere, BlueprintReadWrite, Category = "Dummy
Character")
float BatteryCostPerTick = 5.f;

UPROPERTY(EditAnywhere, BlueprintReadWrite, Category = "Dummy
Character")
float BatteryRechargePerTick = 1.f;

UPROPERTY(EditAnywhere, BlueprintReadWrite, Category = "Dummy
Character")
float RunSpeed = 650.f;

UPROPERTY(EditAnywhere, BlueprintReadWrite, Category = "Dummy
Character")
float WalkSpeed = 500.f;

UPROPERTY(EditAnywhere, BlueprintReadWrite, Category = "Dummy
Character")
float MovementRandomDeviation = 5.f;
```

Here, we have declared a list of variables to handle the agent:

- `MaxBatteryLevel`: Represents the maximum value reachable by the agent battery
- `BatteryCostPerTick`: Represents how much battery power is spent by the agent while it is moving
- `BatteryRechargePerTick`: Represents how much battery power is recovered when the agent is resting
- `RunSpeed`: Represents the maximum speed reachable when running
- `WalkSpeed`: Represents the maximum speed reachable when walking
- `MovementRandomDeviation`: A value that will be randomly added or subtracted to the agent speed in order to make the movement pace less predictable

The UPROPERTY () macro is used to declare class properties with additional functionality and metadata. It allows for easy integration with the Unreal Engine Editor, providing a visual interface to modify and configure these properties. The EditAnywhere property specifier indicates that the property can be edited in property windows from the Unreal Engine Editor, while BlueprintReadWrite specifies that the property will be accessible from the extending Blueprint classes in read and write mode. Finally, we want all the properties to be in the same category – that is, **Dummy Character** – and that's why we have set the Category property specifier.

We need just one more variable, but this need not be public because it will be used for the actor's inner logic. In the protected section, let's add the following declaration:

```
UPROPERTY()
float BatteryLevel;
```

This self-explanatory attribute will be used to keep track of the actual battery level of the agent.

Adding delegates

We now need to create an event dispatcher for battery status change notification; the best method in C++ is to use a **delegate**.

> **Note**
>
> If you are unfamiliar with delegates, my advice is to take a peek at *Appendix A, Understanding C++ in Unreal Engine*, at the end of this book.

Locate the public section and add the following piece of code:

```
DECLARE_DYNAMIC_MULTICAST_DELEGATE_OneParam(FOnBatteryStatusChanged,
EBatteryStatus, NewBatteryStatus);

UPROPERTY(BlueprintAssignable, Category = "Dummy Character")
FOnBatteryStatusChanged OnBatteryStatusChanged;
```

We have declared a dynamic multicast delegate with a single parameter – the battery new status – that will be dispatched every time the battery changes its charge level.

Declaring functions

The last thing we need to add to the header is the function declarations. As a first step, delete the following line of code that is found at the end of the file:

```
virtual void SetupPlayerInputComponent(class UInputComponent*
PlayerInputComponent) override;
```

This character will be controlled by AI, so we don't need to set up the player input. Next, just after the `Tick()` declaration, add the following lines of code:

```
UFUNCTION(BlueprintCallable, Category="Dummy Character")
void SetWalkSpeed();

UFUNCTION(BlueprintCallable, Category="Dummy Character")
void SetRunSpeed();

UFUNCTION(BlueprintCallable, BlueprintGetter, Category="Dummy
Character")
EBatteryStatus GetBatteryStatus() const;
```

We have just declared two functions – `SetWalkSpeed()` and `SetRunSpeed()` – that will let us change the character speed at runtime. Additionally, we have added a getter function for the agent battery status.

In Unreal Engine, the `UFUNCTION()` macro is used to declare functions that are recognized by the Unreal Engine reflection system; this means the function becomes accessible and usable within the Unreal Engine framework. All three functions have the `BlueprintCallable` specifier added, meaning that these functions will be accessible in a Blueprint graph. Additionally, the `GetBatteryStatus()` function has the `const` keyword added; this will remove the execution pin in the corresponding Blueprint node, as we need this function to just be a getter and not change any data during execution.

Implementing functions

Now that all the class declarations have been done, we can start implementing the functions. To do so, the first thing you need to do is to open the `BaseDummyCharacter.cpp` file.

The first thing you'll need to do is to remove the `SetPlayerInputComponent()` function, as the corresponding declaration in the header file was previously removed.

Next, we need to add the `#include` declarations at the very beginning of the file. Simply add these three lines of code:

```
#include "Components/CapsuleComponent.h"
#include "Components/SkeletalMeshComponent.h"
#include "GameFramework/CharacterMovementComponent.h"
#include "BatteryStatus.h"
```

As always, remember to add these `#include` declarations before the `.generated.h` declaration.

Next, locate the `ABaseDummyCharacter()` constructor function, as we will need to set up some character attributes and components. This function should have already a line of code that sets the `bCanEverTick` property to `true`. Add the following line of code just after it:

```
PrimaryActorTick.TickInterval = .25f;
```

As we will use the `Tick()` event just for updating the battery status, we don't need it to be executed every frame; we have set a time interval of a quarter of a second – this will be more than enough to suit our own needs.

Next, add the following lines of code that set up the yaw, pitch, and roll behavior of the character:

```
bUseControllerRotationPitch = false;
bUseControllerRotationYaw = false;
bUseControllerRotationRoll = false;
```

Next, we need to initialize the skeletal mesh component in order to show the dummy puppet model. Add these lines of code:

```
GetMesh()->SetRelativeLocation(FVector(0.f, 0.f, -120.f));
GetMesh()->SetRelativeRotation(FRotator(0.f, -90.f, 0.f));
static ConstructorHelpers::FObjectFinder<USkeletalMesh>
SkeletalMeshAsset(TEXT("/Game/KayKit/PrototypeBits/Character/Dummy.
Dummy"));
if (SkeletalMeshAsset.Succeeded())
{
    GetMesh()->SetSkeletalMesh(SkeletalMeshAsset.Object);
}
GetMesh()->SetAnimationMode(EAnimationMode::AnimationBlueprint);
static ConstructorHelpers::FObjectFinder<UAnimBlueprint>
AnimBlueprintAsset(TEXT("/Game/KayKit/PrototypeBits/Character/ABP_
Dummy.ABP_Dummy"));
if (AnimBlueprintAsset.Succeeded())
{
    GetMesh()->SetAnimClass(AnimBlueprintAsset.Object-
>GeneratedClass);
}
```

Here, we set the mesh location and rotation in order to suit the dummy puppet model. After that, we assign the dummy puppet skeletal mesh asset by hardcoding the asset path; we will use just this asset, so there is no need to assign it from the extending Blueprint classes. We will do the same with the animation Blueprint asset; I have provided one such asset for you in the project files in the declared path.

Now, we are going to set the capsule component size in order to match the dummy puppet model. To do this, add this line of code:

```
GetCapsuleComponent()->InitCapsuleSize(50.f, 120.0f);
```

Finally, set up the movement component by adding the following lines of code:

```
GetCharacterMovement()->bOrientRotationToMovement = true;
GetCharacterMovement()->MaxWalkSpeed = 500.f;
GetCharacterMovement()->RotationRate = FRotator(0.f, 640.f, 0.f);
GetCharacterMovement()->bConstrainToPlane = true;
GetCharacterMovement()->bSnapToPlaneAtStart = true;
GetCharacterMovement()->AvoidanceConsiderationRadius = 2000.f;
GetCharacterMovement()->bUseRVOAvoidance = true;
```

Note the use of `bUseRVOAvoidance`, set to `true`; we will use several agents at the same time, so a basic avoidance system is almost mandatory to make things work properly.

The constructor method is complete, so we can now start implementing all the other functions.

Locate the `BeginPlay()` method, and just after the `Super::BeginPlay()` declaration, add the following lines of code:

```
BatteryLevel = MaxBatteryLevel * FMath::RandRange(0.f, 1.f);
OnBatteryStatusChanged.Broadcast(GetBatteryStatus());
```

When the game starts, we set the AI agent to have a random battery level to make things a bit more interesting, and next, we broadcast this status to all registered listeners.

After that, just after the closing bracket of the `BeginPlay()` function, add the following piece of code:

```
void ABaseDummyCharacter::SetWalkSpeed()
{
    const auto Deviation = FMath::RandRange(-1.f *
MovementRandomDeviation, MovementRandomDeviation);
    GetCharacterMovement()->MaxWalkSpeed = WalkSpeed + Deviation;
}


void ABaseDummyCharacter::SetRunSpeed()
{
    const auto Deviation = FMath::RandRange(-1.f *
MovementRandomDeviation, MovementRandomDeviation);

    GetCharacterMovement()->MaxWalkSpeed = RunSpeed +
MovementRandomDeviation;
}
```

There's nothing fancy here; we just implement the two functions to change the movement speed of the agent, making it walk or run.

We now need to implement the battery status getter function, so add the following lines of code:

```
EBatteryStatus ABaseDummyCharacter::GetBatteryStatus() const
{
    const auto Value = BatteryLevel / MaxBatteryLevel;
    if (Value < 0.05f)
    {
        return EBatteryStatus::EBS_Empty;
    }
    if (Value < 0.35f)
    {
        return EBatteryStatus::EBS_Low;
    }
    if (Value < 0.95f)
    {
        return EBatteryStatus::EBS_Medium;
    }
    return EBatteryStatus::EBS_Full;
}
```

As you can see, we simply check the battery level and return the corresponding status enumeration.

The last thing we need to implement is the `Tick()` function, where we will constantly check how much battery power is consumed, depending on the character movement speed. Locate the `Tick()` function, and just after `Super::Tick(DeltaTime);`, add the following lines of code:

```
const auto CurrentStatus = GetBatteryStatus();
if(GetMovementComponent()->Velocity.Size() > .1f)
{
    BatteryLevel -= BatteryCostPerTick;
}
else
{
    BatteryLevel += BatteryRechargePerTick;
}
BatteryLevel = FMath::Clamp<float>(BatteryLevel, 0.f,
MaxBatteryLevel);
if (const auto NewStatus = GetBatteryStatus();
  CurrentStatus != NewStatus)0
    {
    OnBatteryStatusChanged.Broadcast(NewStatus);
    }
```

In this piece of code, we compute the current battery status using the `GetBatteryStatus()` function. Then, if the velocity of the character's movement is greater than a tiny number – that is, `0.1` – it means the agent is moving, so we decrease the battery level by the `BatteryCostPerTick` value. Otherwise, the agent is standing still – therefore, recharging – so we increase the battery level by `BatteryRechargePerTick`. After that, we clamp the battery level between a value of zero and `MaxBatteryLevel`. Finally, we check whether the starting battery status is different from the new battery status, and we eventually broadcast the new battery status using the `OnBatteryStatusChanged` delegate.

The `BaseDummyCharacter` class has been completed. It is evident that we have not yet incorporated any AI agent behavior; this is intentional, as we plan to manage everything through an `AIController` class, a task we will undertake in the following section.

Creating a behavior tree

In this section, we will create a fully functional behavior tree for the agent we previously created. The steps we will follow are as follows:

- Creating the AI controller
- Creating the Blackboard
- Creating the behavior tree

Let's start by creating a subclass of the `AIController` class to control our dummy puppet.

Creating the AI controller

We will be now creating a class extending `AIController` that will be used as a starting point for the behavior tree. To get started, open the Unreal Engine Editor and do the following steps:

1. From the main menu, select **Tools** | **New C++ Class**.

2. Click on the **All Classes** tab section and look for **AIController**.

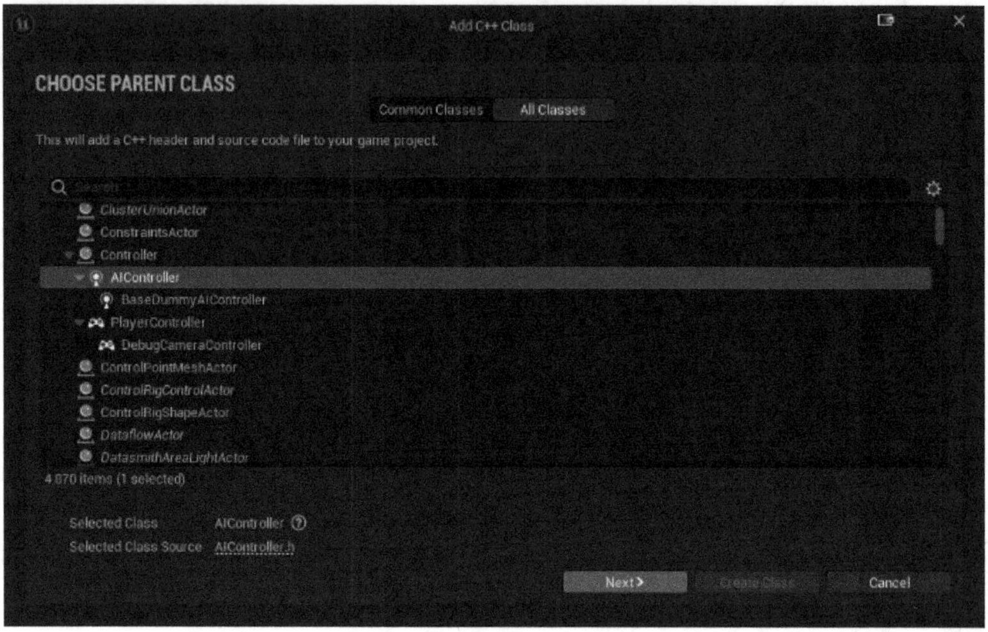

Figure 8.6 – AI controller class creation

3. Click the **Next** button.

4. Name the class `BaseDummyAIController` and click the **Create Class** button.

Once the class files have been created and your IDE has been opened, look for the `BaseDummyAIController.h` header file and open it.

Editing the header file

As a first step, add a forward declaration for the `BehaviorTree` class, just after the `#include` declarations:

```
class UBehaviorTree;
```

Then, in the `protected` section of the header file, add these lines of code:

```
UPROPERTY(EditAnywhere, BlueprintReadOnly,
  Category = "Dummy AI Controller")
TObjectPtr<UBehaviorTree> BehaviorTree;
```

This property declares a pointer to the behavior tree, so that it will be assignable from the **Class Defaults** Blueprints panel – using the EditAnywhere specifier – and it will be readable from any extending Blueprints.

Now, just after these lines of code, add this:

```
virtual void OnPossess(APawn* InPawn) override;
```

The OnPossess() function is called when the controller possesses a Pawn instance – and our dummy character extends it – and it is a good place to run behavior trees.

Implementing the controller

The controller is quite easy to implement; we just need to run the behavior tree when the AI agent is possessed. To do so, open the BaseDummyAIController.cpp file and add these lines of code:

```
void ABaseDummyAIController::OnPossess(APawn* InPawn)
{
    Super::OnPossess(InPawn);

    if (ensureMsgf(BehaviorTree, TEXT("Behavior Tree is nullptr!
Please assign BehaviorTree in your AI Controller.")))
    {
        RunBehaviorTree(BehaviorTree);
    }
}
```

In this function, the Super::OnPossess() base class is called first. Then, we use the ensureMsgf() macro to ensure that the BehaviorTree variable is not a null pointer. If a behavior tree has been set, we run it using the RunBehaviorTree() function.

With the AI controller set, we can start implementing the actual AI behaviors, starting from the Blackboard.

Creating the Blackboard

Creating a Blackboard asset is a straightforward task once you know what keys you will be tracking. In our case, we want to make the following values available to the behavior tree:

- A target location vector that will be used by the agent to walk around the level
- A Boolean flag that will warn the agent once the battery charge is dangerously low
- A Boolean flag that will indicate that the battery has been depleted

To get started, we need to create a Blackboard asset. To do so, follow these steps:

1. Open **Content Drawer** and create a new folder, naming it AI.

2. Open the folder and right-click on **Content Drawer**, selecting **Artificial Intelligence | Blackboard** to create a Blackboard asset.

3. Name the asset BB_Dummy and double-click on it to open it.

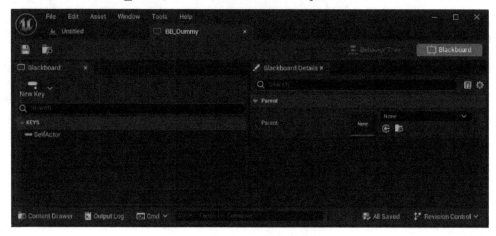

Figure 8.7 – The Blackboard panel

Once the Blackboard is opened, you will see that there's already a key named **SelfActor**; we won't be using it during this chapter, but we will leave it as it is, since it is usually used to store a reference to the owning actor.

We need to create three keys, as stated at the beginning of this subsection, so we'll start by following these steps:

1. Click the **New Key** button, and from the dropdown list, select the **Vector** type, as shown in *Figure 8.8*:

Figure 8.8 – Key creation

2.   Name the new key `TargetLocation`.

3.   Click the **New Key** button again, select the **Bool** type, and name the new key `IsLowOnBattery`.

4.   Click the **New Key** button once more, select the **Bool** type, and name the new key `IsBatteryDepleted`.

Once these steps are finished, your Blackboard should be pretty similar to the one depicted in *Figure 8.9*:

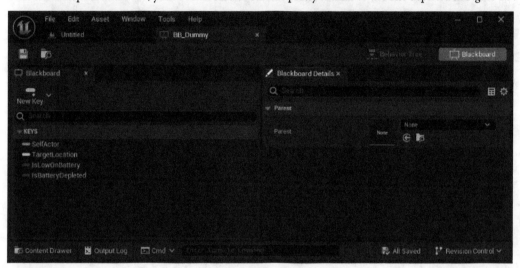

Figure 8.9 – The finished Blackboard

The Blackboard is finished, and we can now start working on the behavior tree for our AI agent.

Creating the behavior tree

Like the Blackboard, a behavior tree is created as a project asset. Let's start by performing the following steps:

1.   In **Content Drawer**, open the `AI` folder.

2.   Right-click on **Content Drawer** and select **Artificial Intelligence | Behavior Tree**.

3.   Name the newly created asset `BT_RoamerDummy` and double-click on it to open it.

Once opened, you should see a graph pretty similar to the one shown in *Figure 8.10*:

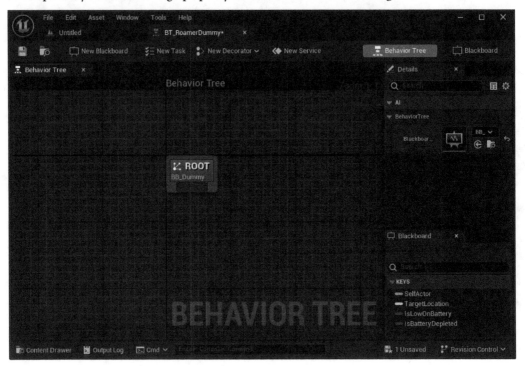

Figure 8.10 – Behavior tree creation

Note that the **Root** node is already included in the graph. As mentioned in the previous chapter, there can only be one such node in the graph. Furthermore, in the **Details** panel, you should see that the **BB_Dummy** Blackboard asset has already been assigned to the Blackboard asset. If you don't see it assigned, simply click the **Blackboard** property dropdown and select the **BB_Dummy** asset; it should be the only asset of that type present in your project.

In this section, we went through a step-by-step guide on creating a character that will utilize a behavior tree to implement AI logic. We have also successfully created a Blackboard asset and a behavior tree asset, which serve as the foundation to implement our character's AI behaviors. As the process of implementing all the AI logic for our character is quite extensive, we will tackle it in the upcoming section. Stay tuned for a detailed walk-through on how to effectively implement AI behaviors for our character.

Implementing behavior tree tasks and services

Before creating the behavior tree, it is crucial to have a clear understanding of the intended implementation. The AI agent will need to manage the following behaviors:

- Wander around the level looking for target locations

- The default movement speed is running, but if the battery level is low, the movement speed should switch to walking

- Once the battery is depleted, it should stop moving

- Once the battery is fully charged, it should return to the movement pattern

It is worth noting that the battery depletion and recharging logic has already been implemented in the `BaseDummyCharacter` class, so we won't need to worry about it – we will just need to listen to the events dispatched by the character and behave accordingly.

As I mentioned in the previous chapter, the Unreal Engine AI system provides a comprehensive collection of tasks, services, and decorators for our AIs. However, it's worth noting that these built-in components may not cover all the specific requirements of your game. After all, as developers, we enjoy the creative process of building new things that suit our unique needs and add to the overall fun of game development!

That's why, before adding nodes to our graph, we will need to implement a set of tasks and services that will make our dummy puppet life easier. In particular, we will create the following:

- A task that will find a random reachable location

- A service that will control the agent speed, depending on the battery level

- A service that will listen for battery status changes

> **Note**
>
> Behavior tree nodes can be implemented both in C++ and in Blueprint; for the purpose of this book, I will be sticking to the C++ option, but in *Chapter 9*, *Extending Behavior Trees*, I will give you some useful hints on how to work also with Blueprint classes.

So, let's start implementing new features for our dummy puppets!

Implementing the FindRandomLocation task

We will begin by implementing the first node, which will enable the AI agent to search for a random actor in the level that has a specific tag. This functionality will allow the agent to have a specific point to navigate toward, enhancing its precision in reaching its destination.

To start implementing this node, from the Unreal Engine Editor, create a new C++ class that extends **BTTaskNode**, and name it BTTask_FindRandomLocation.

Open the BTTask_FindRandomLocation.h header file and add the following declarations:

```
public:
UPROPERTY(EditAnywhere, Category="Blackboard")
FBlackboardKeySelector BlackboardKey;

UPROPERTY(EditAnywhere, Category="Dummy Task")
FName TargetTag;

virtual EBTNodeResult::Type ExecuteTask(UBehaviorTreeComponent&
    OwnerComp, uint8* NodeMemory) override;
```

The BlackboardKey property will be used to declare which key should be used to store the random location once it has been found, while the TargetTag property will be used to find all available actors in the level in order to randomize a selection. Finally, the ExecuteTask() function will be called when the task node needs to be executed and will contain all the logic to randomize a target location. This function will need to return whether the task was successful or failed.

Now, open the BTTask_FindRandomLocation.cpp file, and as a first step, add the needed #include declarations at the top of the file itself:

```
#include "BehaviorTree/BlackboardComponent.h"
#include "Kismet/GameplayStatics.h"
```

Next, add the ExecuteTask() function implementation:

```
EBTNodeResult::Type UBTTask_
FindRandomLocation::ExecuteTask(UBehaviorTreeComponent& OwnerComp,
    uint8* NodeMemory)
{
    const auto BlackboardComp = OwnerComp.GetBlackboardComponent();
    if (BlackboardComp == nullptr) { return EBTNodeResult::Failed; }

    TArray<AActor*> TargetList;
    UGameplayStatics::GetAllActorsWithTag(GetWorld(), TargetTag,
        TargetList);
    if(TargetList.Num() == 0) { return EBTNodeResult::Failed; }

    const auto RandomTarget = TargetList[FMath::RandRange
        (0, TargetList.Num() - 1)];
    BlackboardComp->SetValueAsVector(BlackboardKey.SelectedKeyName,
        RandomTarget->GetActorLocation());
```

```
      return EBTNodeResult::Succeeded;
}
```

The code gets the `Blackboard` component from `OwnerComp` and checks its validity. Then, it retrieves a list of actors with a specific tag, selecting a random element from that list. Then, it updates the Blackboard with the selected target's location by using the `SetValueAsVector()` method. Note the use of `EBTNodeResult::Succeeded` and `EBTNodeResult::Failed` to return the result of all these operations; this is a requirement in order to indicate to the behavior tree whether a task was successful or not.

Now that we have completed this task node, we can move on to the next step, which involves creating a service.

Implementing the SpeedControl service

We are now prepared to create our first custom service, which will monitor the character's speed based on the battery charge. As you may remember from the previous chapter, a service is usually run at fixed intervals, and that's exactly what we will do for the speed control service class – we will check the battery status and change the character speed accordingly.

To start implementing this class, from the Unreal Engine Editor, create a new C++ class that extends **BTService**, and name it `BTService_SpeedControl`.

Open the `BTTask_SpeedControl.h` header file, and in the `public` section, add the following declarations:

```
protected:
  virtual void TickNode(UBehaviorTreeComponent& OwnerComp, uint8*
    NodeMemory, float DeltaSeconds) override;
```

The `TickNode()` function that we are overriding will be executed on every node tick interval that this service is attached to. In order to call this function, the `bNotifyTick` needs to be set to `true`; this value is already set by default, but it's good to know if you need to disable it – something we'll implement in the next service.

We are ready to implement the service, so open the `BT_Service_SpeedControl.cpp` file and add the following `#include` declarations at the top:

```
#include "BaseDummyAIController.h"
#include "BaseDummyCharacter.h"
```

After that, add the `TickNode()` implementation:

```cpp
void UBTService_SpeedControl::TickNode(UBehaviorTreeComponent&
  OwnerComp, uint8* NodeMemory, float DeltaSeconds)
{
    Super::TickNode(OwnerComp, NodeMemory, DeltaSeconds);

    const auto AIController = Cast<ABaseDummyAIController>(OwnerComp.
      GetAIOwner());
    if (!AIController) return;

    const auto ControlledCharacter = Cast<ABaseDummyCharacter>
      (AIController->GetPawn());
    if (!ControlledCharacter) return;

    switch (ControlledCharacter->GetBatteryStatus())
    {
      case EBatteryStatus::EBS_Empty:
          break;
      case EBatteryStatus::EBS_Low:
          ControlledCharacter->SetWalkSpeed();
          break;
      case EBatteryStatus::EBS_Medium:
      case EBatteryStatus::EBS_Full:
          ControlledCharacter->SetRunSpeed();
          break;
    }
}
```

This function is quite straightforward; all it does is update the speed of the controlled character based on the character's battery status.

At this stage, you might be wondering why we retrieve the AI controller and character reference at every tick instead of storing them. It may seem inefficient in terms of computational power, but it's important to remember that the behavior tree is a shared asset. This means that a single instance of the behavior tree – and its nodes – will be executed for all AI agents that use it. Hence, storing a class reference would not yield any advantages and could result in unpredictable behaviors.

If there is a genuine need to store a reference, you would need to create a node instance, which is precisely what we will do with the upcoming service.

Implementing the BatteryCheck service

The second service we are about to create is going to be a bit more challenging than the previous one. We need to continuously monitor any changes in the battery status; the most straightforward approach would be to use the node tick to constantly check the character's battery status, similar to how the `UBTService_SpeedControl` class operates. However, as we learned earlier in this chapter, the dummy character dispatches battery status events. So, why not leverage this feature and make use of it?

Let's start implementing this service; from the Unreal Engine Editor, create a new C++ class that extends **BTService** and name it `BTService_BetteryCheck`.

Once the files have been created, open the `BTService_BatteryCheck.h` file, and just after the `#include` section, add the following forward declarations:

```
class ABaseDummyCharacter;
enum class EBatteryStatus : uint8;
```

Then, add the `public` section and the constructor declaration:

```
public:
UBTService_BatteryCheck();
```

Just after that, declare the `protected` section, along with all the necessary properties:

```
protected:

UPROPERTY()
UBlackboardComponent* BlackboardComponent = nullptr;

UPROPERTY()
ABaseDummyCharacter* ControlledCharacter = nullptr;

UPROPERTY(BlueprintReadOnly, EditAnywhere, Category="Blackboard")
FBlackboardKeySelector IsLowOnBatteryKey;

UPROPERTY(BlueprintReadOnly, EditAnywhere, Category="Blackboard")
FBlackboardKeySelector IsBatteryDepletedKey;
```

As you can see, we are doing something slightly different from the previous service; we are declaring a reference to the `Blackboard` component and the character. In this case, we will work with node instances, so each AI agent will have its own separate instance of the node decorated by this service. We also declare two Blackboard keys to assign the proper values to the Blackboard.

Just after that, add the following functions:

```
virtual void OnBecomeRelevant(UBehaviorTreeComponent& OwnerComp,
uint8* NodeMemory) override;

virtual void OnCeaseRelevant(UBehaviorTreeComponent& OwnerComp, uint8*
NodeMemory) override;

UFUNCTION()
void OnBatteryStatusChange(EBatteryStatus NewBatteryStatus);
```

The `OnBecomeRelevant()` function will be called when the decorated node becomes active, while the `OnCeaseRelevant()` function is no longer active. We will use these two functions to register the battery status events and react accordingly by using the `OnBatteryStatusChange()` function.

You can now open the `BTService_BatteryCheck.cpp` file to start implementing the functions. As a first step, add the needed `#include` declarations at the top of the file:

```
#include "BaseDummyAIController.h"
#include "BaseDummyCharacter.h"
#include "BatteryStatus.h"
#include "BehaviorTree/BlackboardComponent.h"
```

Immediately after that, add the constructor implementation:

```
UBTService_BatteryCheck::UBTService_BatteryCheck()
{
    bCreateNodeInstance = true;
    bNotifyBecomeRelevant = true;
    bNotifyCeaseRelevant = true;
    bNotifyTick = false;
}
```

Although it may appear that we are simply setting some flags, we are actually making a significant change to the behavior of this service. First, we create a node instance; each AI agent will have its own instance of this service. Next, we disable the service tick, as we won't be needing it, and we activate the relevancy behavior.

Next, let's implement the `OnBatteryStatusChange()` method:

```
void UBTService_BatteryCheck::OnBatteryStatusChange(const
EBatteryStatus NewBatteryStatus)
{
    switch (NewBatteryStatus)
    {
    case EBatteryStatus::EBS_Empty:
```

```
            BlackboardComponent->SetValueAsBool(IsBatteryDepletedKey.
                SelectedKeyName, true);
            break;
        case EBatteryStatus::EBS_Low:
            BlackboardComponent->SetValueAsBool(IsLowOnBatteryKey.
                SelectedKeyName, true);
            BlackboardComponent->SetValueAsBool(IsBatteryDepletedKey.
                SelectedKeyName, false);
            break;
        case EBatteryStatus::EBS_Medium:
            break;
        case EBatteryStatus::EBS_Full:
            BlackboardComponent->SetValueAsBool(IsLowOnBatteryKey.
                SelectedKeyName, false);
            break;
    }
}
```

There is nothing fancy here; we just set the Blackboard keys based on the new battery status. After that, we implement the two remaining functions:

```
void UBTService_BatteryCheck::OnBecomeRelevant(UBehaviorTreeComponent&
OwnerComp, uint8* NodeMemory)
{
    Super::OnBecomeRelevant(OwnerComp, NodeMemory);

    BlackboardComponent = OwnerComp.GetBlackboardComponent();

    const ABaseDummyAIController* AIController =
      Cast<ABaseDummyAIController>(OwnerComp.GetAIOwner());
    if (!AIController) return;
    APawn* ControlledPawn = AIController->GetPawn();
    if (!ControlledPawn) return;
    ControlledCharacter = Cast<ABaseDummyCharacter>(ControlledPawn);
    if (!ControlledCharacter) return;

    ControlledCharacter->OnBatteryStatusChanged.AddDynamic
      (this, &UBTService_BatteryCheck::OnBatteryStatusChange);
}

void UBTService_BatteryCheck::OnCeaseRelevant(UBehaviorTreeComponent&
  OwnerComp, uint8* NodeMemory)
{
    Super::OnCeaseRelevant(OwnerComp, NodeMemory);
```

```
        ControlledCharacter->OnBatteryStatusChanged.RemoveDynamic
            (this, &UBTService_BatteryCheck::OnBatteryStatusChange);
    }
```

These two functions register and unregister the OnBatteryStatusChanged delegate; additionally, OnBecomeRelevantFunction() saves a reference to the Blackboard component and the AI controller – something we can do because we use an instanced node for this service.

In this extensive section, you acquired the knowledge to create custom tasks and services in C++. Often, the pre-built classes provided by Unreal Engine may not suffice to create engaging AI behaviors. Hence, it becomes essential to develop your own distinctive nodes. In the forthcoming section, we will create such new classes to construct a fully operational AI agent.

Setting up a behavior tree on an agent

To begin delving into the AI agent behavior tree, the first step is to compile the entire project. Once the process is finished, your custom tasks and services will be available as an option in the behavior tree.

> **Note**
>
> If you are unfamiliar with the Unreal Engine compilation process, my advice is to take a peek at *Appendix A, Understanding C++ in Unreal Engine*, and then return to this chapter.

Once the compilation phase is finished, we can start editing the behavior tree.

Editing the behavior tree

Open the **BT_RoamerDummy** asset we previously created and locate the only element – the **ROOT** node – present in the graph; you will see that it has a darker area at the bottom.

Figure 8.11 – The ROOT node

Clicking and dragging from this area will make all the nodes that can be connected to the **ROOT** node available.

Note

Moving forward, whenever I mention the task of adding a node, I will be requesting that you perform the aforementioned action. One such case is depicted in *Figure 8.12*.

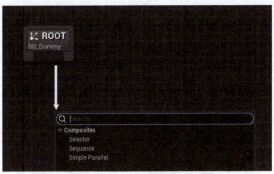

Figure 8.12 – Adding a node

To get started, do the following steps:

1. Add a **Composites | Sequence** node to the **Root** node, and in the **Details** panel, rename it Root Sequence.

2. From the **Root Sequence** node, add a **Tasks | FindRandomLocation** node.

3. With the newly created node selected, in the **Details** panel, set the **Blackboard Key** dropdown value to **TargetLocation**.

4. From the **Root Sequence** node, add a **Composites | Selector** node at the right of the **FindRandomLocation** node.

    Your graph should now be similar to the one depicted in *Figure 8.13*:

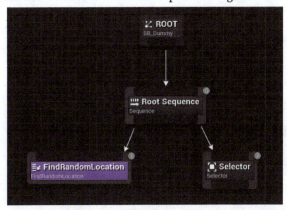

Figure 8.13 – The initial behavior tree

We now need to decorate the **Selector** node to give it extra abilities. To do so, follow these steps:

5. Right-click on the **Selector** node and select **Add Decorator | Conditional Loop**.

6. Right-click one more time and select **Add Service | Battery Check**.

7. Click on the **Conditional Loop** decorator, and in the **Details** panel, set the **Blackboard Key** attribute dropdown to **TargetLocation**. The **Key Query** attribute should be left to **Is Set**.

8. Click on the **BatteryCheck** service, and in the **Details** panel, do the following:

    - Set the **Is Low on Battery Key** dropdown value to **IsLowOnBattery**.

    - Set the **Is Battery Depleted Key** dropdown value to **IsBatteryDepleted**.

The **Selector** node should now look similar to *Figure 8.14*:

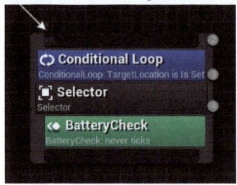

Figure 8.14 – The decorated Selector node

Note that the **BatteryCheck** service should show a **never ticks** comment; this is what we set when we implemented the C++ class.

What we have done so far is basically the main loop of the AI agent behavior; we start by finding a target location for the AI agent, and then we execute a **Selector** node that will be waiting for battery status notifications. The **Conditional Loop** decorator will keep on repeating the child nodes (still to be added) while the **TargetLocation** key is set.

Now, we will focus on the **Selector** node and do the following:

9. Add a **Composite | Sequence** child node and rename it `Roam Sequence`.

10. Add a **Task | Wait** child node, and in its **Details** panel, do the following:

    - Set the **Wait Time** value to `8.0`

    - Set the **Random Deviation** value to `2.0`

    - Set the **Node Name** as `Recharge Battery`

We need to decorate the **Roam Sequence** node with extra capabilities, so we do the following:

11. Right-click on the **Roam Sequence** node and select **Add Decorator | Blackboard**.

12. With the decorator selected, do the following in the **Details** panel:

- Set the **Notify Observers** dropdown value to **On Value Change**

- Set the **Observer aborts** dropdown value to **Self**

- Set the **Key Query** dropdown value to **Is Not Set**

- Set the **Blackboard Key** dropdown value to **IsBatteryDepleted**

This portion of the graph should now look like *Figure 8.15*:

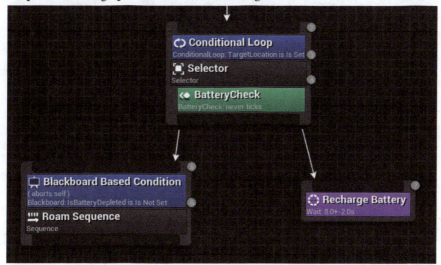

Figure 8.15 – The roam loop

This portion of the graph will constantly loop between two phases – a roam sequence and a **Wait** node. The AI agent will stay in the roam sequence until the battery has been depleted. After that, it will stay still for between 6 and 10 seconds (so that the battery will recharge) and then revert to roaming.

The last step we need to take is to implement the roaming nodes. To do this, from the **Roam Sequence** node, do the following:

13. Add a **Tasks | Move To** node, and in the **Details** panel, set the **Blackboard Key** dropdown value to **TargetLocation**.

14. Add a **Tasks | FindRandomLocation** node and, in the **Details** panel, set the **Target Tag** value to TargetPoint.

15. Add a **Tasks | Wait** node, leaving its properties to their default values.

16. We need to add extra functionality to the **Move To** node, so right-click on it and select **Add Service | Speed Control**. This portion of the graph should look like *Figure 8.16*:

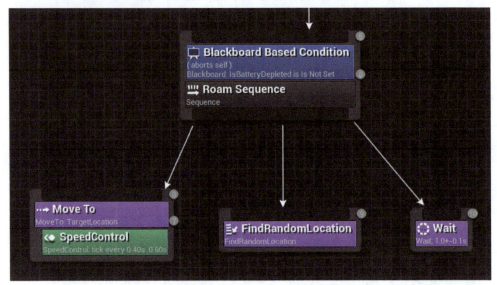

Figure 8.16 – The roam sequence

What we are doing here is pretty straightforward; we try to reach the target point and, once reached, search for another one, and then get some well-deserved rest by waiting a random time interval. While moving, we constantly check the battery status, changing the speed of the AI agent accordingly.

Great news! The behavior tree graph has been completed. Now, we can attach it to our dummy puppet and witness its behavior. However, before we proceed, we need to create some suitable Blueprints to ensure that everything functions smoothly.

Creating the AI agent Blueprints

When working with Unreal Engine, it is considered good practice to create a Blueprint from a C++ class. This approach offers several advantages, such as flexibility and extensibility, that contribute to a more efficient development process. That's why we are going to create some Blueprints from our previously created classes.

Creating the controller Blueprint

Let's start by extending the `BaseDummyAIController` class into a Blueprint. To do so, in your **Content Browser**, create a new folder, name it `Blueprints`, and then follow these steps:

1. Create a new Blueprint class deriving from **BaseDummyAIController**, and name it `AIRoamerDummyController`.

2. Open it, and in the **Class Defaults** panel, look for the **Dummy AI Controller** category and set the **Behavior Tree** attribute to **BT_RoamerDummy**.

Figure 8.17 – Assigning the behavior tree

This is all you need to do to set up the AI controller for the roamer dummy character; we will now create the dedicated character.

Creating the character Blueprint

To create a character Blueprint, follow these steps:

1. In the `Blueprints` folder of your **Content Browser** create a new Blueprint class deriving from **BaseDummyCharacter**, and name it `BP_RoamerDummyCharacter`.

2. Open it, and in the **Class Defaults** panel, look for the **AI Controller Class** attribute located in the **Pawn** category. From the dropdown menu, select **AIRoamerDummyController**.

Congratulations on creating your very own roamer agent! Now, we just need to add one final touch to make it even better – a battery indicator.

Adding cosmetics

To provide visual feedback on the AI agent's status, we will create a component that displays the battery charge level through a light. This visual indicator will adjust the intensity of the light based on the current charge level. The higher the charge, the more intense the light will be. This will allow users to easily gauge the agent's battery status at a glance, enhancing the overall user experience and ensuring they are aware of the agent's power level.

Let's start by creating a new C++ class that inherits from the `StaticMeshComponent` class, calling it `BatteryInjdicatorComponent`. Once the class has been created, open the `BatteryIndicatorComponent.h` file and replace the `UCLASS()` line with the following one:

```
UCLASS(BlueprintType, Blueprintable, ClassGroup="Unreal Agility
Arena", meta=(BlueprintSpawnableComponent))
```

This will make the component accessible to Blueprints. Then, after the `GENERATED_BODY()` line of code, add the following code:

```
public:
UBatteryIndicatorComponent();
```

```
protected:
UPROPERTY()
UMaterialInstanceDynamic* DynamicMaterialInstance;

virtual void BeginPlay() override;

UFUNCTION()
void OnBatteryStatusChange(EBatteryStatus NewBatteryStatus);
```

The only things that need an explanation here are the DynamicMaterialInstance property, which will be used to change the material intensity property to make it more or less bright, and the OnBatteryStatusChange() function, which will be used to handle battery status change events.

Now, to start implementing the component, open the BatteryIndicatorComponent.cpp file, and add the following declaration at the top of it:

```
#include "BaseDummyCharacter.h"
```

The constructor just needs to set the static mesh asset, so add this piece of code:

```
UBatteryIndicatorComponent::UBatteryIndicatorComponent()
{
    static ConstructorHelpers::FObjectFinder<UStaticMesh>
      StaticMeshAsset(
      TEXT("/Game/_GENERATED/MarcoSecchi/SM_HeadLight.SM_
        Headlight"));
    if (StaticMeshAsset.Succeeded())
    {
        UStaticMeshComponent::SetStaticMesh(StaticMeshAsset.Object);
    }
}
```

Next, implement the BeginPlay() function by adding this block of code:

```
void UBatteryIndicatorComponent::BeginPlay()
{
    Super::BeginPlay();

    ABaseDummyCharacter* Owner =
      Cast<ABaseDummyCharacter>(GetOwner());
    if(Owner == nullptr) return;

    AttachToComponent(Owner->GetMesh(),
      FAttachmentTransformRules::SnapToTargetIncludingScale,
```

```
"helmet");

    DynamicMaterialInstance = this->CreateDynamicMaterialInstance
        (1, GetMaterial(1));

    Owner->OnBatteryStatusChanged.AddDynamic
        (this, &UBatteryIndicatorComponent::OnBatteryStatusChange);
}
```

In this function, we check whether the owner of this component is an instance of the BaseDummyCharacter class. Next, we attach this component to the owner's mesh component on a socket named **helmet** – a socket I already provided for you in the dummy skeletal mesh, as you can see in *Figure 8.18*:

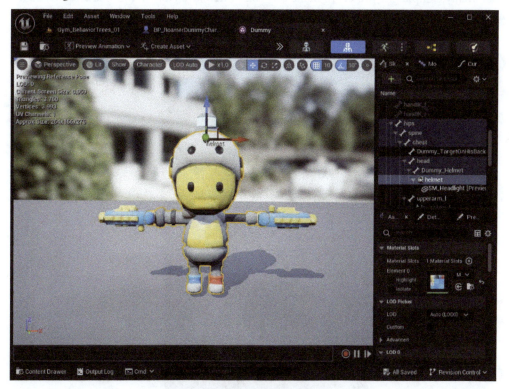

Figure 8.18 – The helmet socket

After that, we create a dynamic material instance for this component – this will let us modify the material properties at runtime. Finally, we add an event handler to the OnBatteryStatusChanged event of the owner character, which calls the OnBatteryStatusChange function of this component whenever the battery status of the owner character changes.

With this function complete, we just need to add the event handler to our code:

```
void UBatteryIndicatorComponent::OnBatteryStatusChange(EBatteryStatus
NewBatteryStatus)
{
    const auto BatteryValue = StaticCast<float>(NewBatteryStatus);
    const auto Intensity = (BatteryValue - 1.f) * 25.f;
    DynamicMaterialInstance->SetScalarParameterValue(FName
        ("Intensity"), Intensity);
}
```

Here, we convert the `NewBatteryStatus` enum to a `float` value, calculate the light intensity, and then set a scalar parameter, `Intensity`, in the dynamic material instance.

We can finally compile the project to make this component available to the character. Once the compilation process is finished, open the **BP_RoamerDummy** Blueprint and do the following:

1.   In the **Components** panel, click the + **Add** button.

2.   Select **UnrealAgilityArena | Battery Indicator** to add this component to the character.

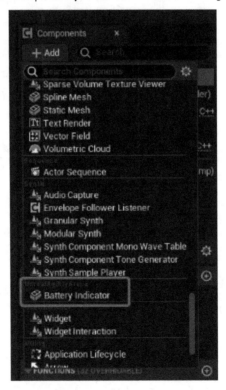

Figure 8.19 – Adding a component

Your dummy character is all set and eagerly awaiting to be tested at a suitable gym level.

Testing an agent in a gym

We are now ready to create a gym for our AI agent and see how it behaves. As you already know from *Chapter 4, Setting Up a Navigation Mesh*, how to properly set up a gym with a nav mesh, I won't dive into the creation details. Instead, I will give you some generic info about the level creation. Here is what you should do:

1. Create a level of your choice, starting from the Level Instances and Packed Level Actors I provided in the project template

2. Add a **NavMeshBoundsVolume** actor so that it will cover all the walkable areas

3. Add some obstacles to make things more interesting

4. Add one or more **BP_RoamerDummyCharacter** instances

5. Add some **NS_Target** Niagara actors that will work as target points

The only thing worth noting is that your AI agent will look for target points with the **TargetPoint** tag. If you are unfamiliar with the tag system, here's how to tag your actors:

1. For each **NS_Target** Niagara system, search for the **Tags** attribute located in the **Details** panel.

2. Click the + button to add a new tag.

3. In the **Index [0]** field that will be created, insert TargetPoint, as shown in *Figure 8.20*.

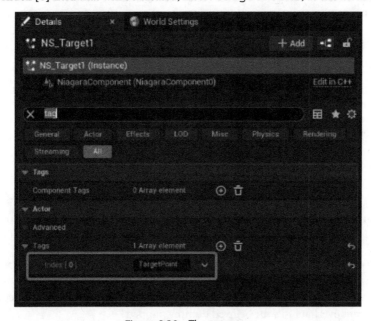

Figure 8.20 – The actor tag

Once your level is finished, you are ready to start testing it; mine is shown in *Figure 8.21*:

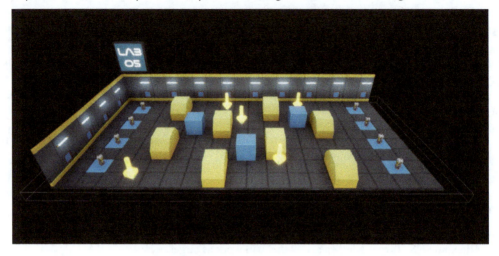

Figure 8.21 – The finished level

Upon starting the simulation, you should see the following things happen:

- Depending on the starting battery charge (which is randomized), your AI agents will start to run, walk, or stand still

- They will try to reach a target point, and once reached, they will rest for about a second and then look for another one

- If the battery charge is low, they will start walking instead of running

- If the battery charge is depleted, they will stop, start recharging, and then start running again

- The headlight should be brighter when the AI agent is fully charged and turn off when the battery is low

That ends this section, where you have learned how to build a fully working behavior tree, including creating your customized tasks and services.

Summary

In this quite lengthy chapter, you were introduced to the basics of creating a behavior tree in Unreal Engine. As you have learned, creating a fully working AI agent is a mixture of out-of-the-box features, custom classes, and a bit of ingenuity.

We're just getting started on a fascinating journey that will unfold in the upcoming chapters, starting with the next one, where we'll dive into the intricate workings of tasks, services, and decorators. Get ready for a significant overhaul to your beloved dummy puppets!

9

Extending Behavior Trees

Understanding the nuances of behavior trees is crucial for game developers, as they enable them to have a firm grasp on creating responsive and more engaging AI characters. This is why, in this chapter, we will embark on a deep dive into the inner workings and best practices of the behavior tree system for Unreal Engine; we will go through a thorough explanation of their functionality, most notably on how to create more complex custom tasks, services, and decorators for our AI agents. Additionally, we will go back to the debugging tools and see how to analyze behavior trees at runtime.

In this chapter, we will be covering the following topics:

- Presenting best practices for authoring behavior trees
- Understanding decorators
- Understanding services
- Understanding tasks
- Debugging behavior trees

Technical requirements

To follow the topics presented in this chapter, you should have completed the previous ones and understood their content.

Additionally, if you would prefer to begin with code from the companion repository for this book, you can download the `.zip` project files provided in this book's companion project repository: `https://github.com/PacktPublishing/Artificial-Intelligence-in-Unreal-Engine-5`.

To download the files from the end of the last chapter, click the `Unreal Agility Arena - Chapter 08 - End` link.

Presenting best practices for authoring behavior trees

It seems our little novel has a new chapter:

In their quest to test the capabilities of their AI puppets, Dr. Markus and Professor Viktoria decided to equip them with Nerf guns. The idea was to create a fun and engaging scenario where the puppets could showcase their newfound shooting skills.

In the secret lab, the atmosphere buzzed with excitement as the puppets, now armed and ready with non-lethal weapons, were going to take on the challenge. With their advanced AI programming and sensor systems, the puppets were able to analyze the environment, calculate trajectories, and take aim at the targets with remarkable accuracy.

When designing a behavior tree for your AI agents, it is crucial to possess a thorough understanding of the best practices to consider. Additionally, staying up to date with the latest advancements and research in the field of AI in the video game industry will provide valuable insights and inform your decision-making process when designing better behavior trees.

Listing best practices

In this subsection, I will provide you with some valuable advice to optimize and enhance the engagement of your AIs; some of them come from well-known patterns, while some originate from my own personal experience – most of the time, acquired by a trial-and-error process.

Use proper naming conventions

It is good practice to properly name any newly created tasks, decorators, or services in the behavior tree. Use naming conventions that indicate the type of asset, be it a task, decorator, or service. As a rule of thumb, you should use the following prefixes:

- `BTTask_` for tasks
- `BTDecorator_` for decorators
- `BTService_` for services

This will not only help you in making it clear the type of asset you have created but you will also get a beneficial side effect; in the behavior tree graph, the system will recognize the type of class and will remove the prefix by showing just the node name. The use of improper naming conventions would result in erratic behaviors when trying to select the nodes in the node graph.

As an example, in *Chapter 8, Setting Up a Behavior Tree*, you may have noticed that we named the battery check service `BTService_BatteryCheck`, but in the behavior tree graph, it is shown just as **Battery Check**.

Give your nodes a meaningful name

Behavior tree tasks, decorators, and services all have a property named `NodeName` that will be used to display the node name in the graph; use it if you want to give your nodes a different name than the class name.

Avoid directly changing node properties

Instead of directly changing the properties of nodes within the behavior tree, you should leverage the power of Blackboards and change its keys. Alternatively, you may call a function inside the character that makes the necessary modifications to keys. This will help maintain a cleaner and more organized structure.

As an example, our previously created dummy puppet handles the battery logic through the character and some dedicated Blackboard keys.

Consider performance optimization

The behavior tree system in Unreal Engine is fundamentally optimized, as it avoids evaluating the entire tree at every frame and instead relies on success and fail notifications to determine the next node to execute. However, it is important to exercise caution and not rely too heavily on certain features with the assumption that they will automatically function as intended.

As an example, in the *Battery Check* service we previously created, we disabled the tick interval, leveraging the power of delegates instead.

Use modular design

Try breaking down complex behaviors into smaller, reusable modules such as tasks, services, and decorators. This modular approach will make it easier to maintain and update the behavior tree.

As an example of small tasks creating a more complex behavior, try checking the **Roam Sequence** from the behavior tree implemented in *Chapter 8*, *Setting Up a Behavior Tree*.

Do not assume your character is going to be an AI agent

When developing an AI character, you may be tempted to add AI logic directly into the `Character` class. However, it is generally advisable to avoid this practice. The `AI Controller` class has been specifically designed to serve a purpose, offering a more structured and modular method for overseeing the AI behavior of characters. By leveraging the `AI Controller` class, you can effectively separate the AI logic from the character itself, resulting in enhanced maintainability and the ability to update the AI behavior independently.

As an example, having the capability of switching between an AI controller and a player controller at runtime can provide several advantages such as allowing a player to take control of an AI character.

As a rule of thumb, it is recommended to create your `Pawn` and `Character` classes in a way that allows them to be controlled by both players and AI interchangeably.

Later in this chapter, we will be creating a brand-new AI agent with this in mind.

Debug and test often

Regularly test and debug your behavior tree to ensure it functions as intended. Use debugging tools provided by Unreal Engine to identify and resolve any issues.

By the end of this chapter, I will be showing you how to properly use the debugging tools with your behavior trees.

Now that you have some additional information on how to create an AI agent effectively, it's time to dive back into our project and begin crafting a brand-new character! Let's get started!

Implementing a gunner character logic

As a starting point, we will be creating a new AI character with some extra features; in particular, it will have some nice gunning capabilities to shoot at targets – by means of a non-lethal Nerf gun. As mentioned earlier, it's important to prioritize simplicity and modularity when working on your characters. In this particular scenario, it doesn't make sense to have the shooting logic integrated within the base character. Instead, it would be more effective to take advantage of the capabilities provided by Unreal Engine components. By doing so, you can keep the shooting functionality separate and modular, allowing for better organization and flexibility in your project. As a final result, the AI agent will be capable of shooting at targets, by using AI behaviors.

Creating the BaseWeaponComponent class

To create a weapon component for our dummy character, we will start by extending a `StaticMeshComponent` class; this will give a good starting point – the mesh – and we will just need to add the socket attaching logic and the shooting logic.

To start creating this component, from Unreal Engine, create a new C++ class extending from `StaticMeshComponent` and name it `BaseWeaponComponent`.

Once the class has been created, open the `BaseWeaponComponent.h` file and change the `UCLASS()` macro with this line of code:

```
UCLASS(BlueprintType, Blueprintable, ClassGroup="UnrealAgilityArena",
    meta=(BlueprintSpawnableComponent))
```

This will make the component accessible to Blueprints, and you can attach it directly to a Blueprint class.

Now, inside the class, just after the GENERATED_BODY() macro, add the following declarations:

```
public:
    UBaseWeaponComponent();

    UFUNCTION(BlueprintCallable)
    virtual void Shoot();

    UPROPERTY(EditAnywhere, BlueprintReadWrite, Category="Bullet")
    TSubclassOf<AActor> BulletClass;

    UPROPERTY(EditAnywhere, BlueprintReadWrite, Category="Bullet")
    FVector MuzzleOffset = FVector(150, 30.f, 0.f);

protected:
    virtual void BeginPlay() override;
```

Here, we are declaring the constructor, a Shoot() function that will spawn a bullet; then, we declare the BulletClass property for the spawned bullet and MuzzleOffset to precisely place the bullet spawn point. Finally, in the protected section, we will need the BeginPlay() declaration to add some initialization.

We are now ready to implement the component, so open the BaseWeaponComponent.cpp file and, at the very top of it, add this line of code:

```
#include "BaseDummyCharacter.h"
```

Immediately after that, add the class constructor:

```
UBaseWeaponComponent::UBaseWeaponComponent()
{
    static ConstructorHelpers::FObjectFinder<UStaticMesh>
      StaticMeshAsset(
        TEXT("/Game/KayKit/PrototypeBits/Models/Gun_Pistol.Gun_
          Pistol"));
    if (StaticMeshAsset.Succeeded())
    {
        UStaticMeshComponent::SetStaticMesh(StaticMeshAsset.Object);
    }
}
```

This function is quite easy and straightforward as we are just declaring a default mesh – a Nerf gun pistol – for the component; you will be free to change it later on when extending this class with a Blueprint.

After that, add the `BeginPlay()` implementation:

```
void UBaseWeaponComponent::BeginPlay()
{
    Super::BeginPlay();
    const auto Character = Cast<ABaseDummyCharacter>(GetOwner());
    if(Character == nullptr) return;

    AttachToComponent(Character->GetMesh(),
        FAttachmentTransformRules::SnapToTargetIncludingScale,
        "hand_right");
}
```

This function tries to cast the owner of the component to a `Character` object, and if the cast is successful, it attaches the component to the mesh of the character at the `hand_right` socket so that the gun will stay in the character's right hand; I have already provided such a socket for you in the **Dummy** skeletal mesh, as shown in *Figure 9.1*:

Figure 9.1 – Hand socket

As a final step, it's time to implement the Shoot() function logic, so add the following lines of code:

```
void UBaseWeaponComponent::Shoot()
{
    if (BulletClass == nullptr) return;

    auto const World = GetWorld();

    if (World == nullptr) return;

    const FRotator SpawnRotation = GetOwner()->GetActorRotation();

    const FVector SpawnLocation = GetOwner()->GetActorLocation() +
      SpawnRotation.RotateVector(MuzzleOffset);

    FActorSpawnParameters ActorSpawnParams;
    ActorSpawnParams.SpawnCollisionHandlingOverride =
      ESpawnActorCollisionHandlingMethod::AdjustIfPossibleButDont
      SpawnIfColliding;

    World->SpawnActor<AActor>(BulletClass, SpawnLocation,
SpawnRotation, ActorSpawnParams);
}
```

This function checks whether BulletClass is valid, gets the reference to World – the top-level object that represents a map reference – then calculates the spawn location and rotation based on the owner's position, and finally, spawns an actor using BulletClass at the calculated location and rotation.

The component is now ready; the next step will be to create a proper bullet to shoot at the right time.

Creating the BaseBullet class

Having created a weapon component that spawns bullets, the next logical step is to create a spawnable bullet. This is going to be quite straightforward; we will be creating an object with a mesh that will move forward, doing damage to anything that it will hit. Let's start by creating a new C++ class extending Actor and calling it BaseGunBullet; after the class has been created, open the BaseGunBullet.h header file and, just after the #include section, add the following forward declarations:

```
class USphereComponent;
class UProjectileMovementComponent;
class UStaticMeshComponent;
```

After that, change the UCLASS() macro so that this class is an acceptable base for creating Blueprints:

```
UCLASS(Blueprintable)
```

Now, just after the `GENERATED_BODY()` macro, add the needed component declarations:

```
UPROPERTY(VisibleDefaultsOnly, Category="Projectile")
USphereComponent* CollisionComponent;

UPROPERTY(VisibleAnywhere, BlueprintReadOnly, Category="Projectile",
  meta=(AllowPrivateAccess="true"))
UStaticMeshComponent* MeshComponent;

UPROPERTY(VisibleAnywhere, BlueprintReadOnly, Category="Movement",
  meta=(AllowPrivateAccess="true"))
UProjectileMovementComponent* ProjectileMovementComponent;
```

The next step is to declare the `public` functions – that is, the constructor and the getters for the components – so add the following lines of code:

```
public:
    ABaseGunBullet();
    USphereComponent* GetCollision() const { return
      CollisionComponent; }
    UProjectileMovementComponent* GetProjectileMovement() const {
      return ProjectileMovementComponent; }
    UStaticMeshComponent* GetMesh() const { return MeshComponent; }
```

In the `protected` section, remove the `BeginPlay()` declaration as it won't be needed. Instead, we will need the `OnHit()` handler for bullet collision events:

```
protected:
    virtual void BeginPlay() override;

    UFUNCTION()
    void OnHit(UPrimitiveComponent* HitComp, AActor* OtherActor,
      UPrimitiveComponent* OtherComp, FVector NormalImpulse, const
        FHitResult& Hit);
```

Now that the header is complete, it's time to start implementing the class, so open up `BaseGunBullet.cpp`. As a first step, add the needed `#include` declarations at the top of the file:

```
#include "GameFramework/ProjectileMovementComponent.h"
#include "Components/SphereComponent.h"
#include "Components/StaticMeshComponent.h"
#include "Engine/DamageEvents.h"
```

Next, remove the `BeginPlay()` implementation, which, as I previously said, won't be needed. After that, add the constructor implementation:

```
ABaseGunBullet::ABaseGunBullet()
{
    PrimaryActorTick.bCanEverTick = false;
    InitialLifeSpan = 10.0f;

    CollisionComponent =
      CreateDefaultSubobject<USphereComponent>(TEXT("Collision"));
    CollisionComponent->InitSphereRadius(20.0f);
    CollisionComponent->BodyInstance.
      SetCollisionProfileName("BlockAll");
    CollisionComponent->OnComponentHit.AddDynamic(this,
      &ABaseGunBullet::OnHit);
    CollisionComponent->SetWalkableSlopeOverride
      (FWalkableSlopeOverride(WalkableSlope_Unwalkable, 0.f));
    CollisionComponent->CanCharacterStepUpOn = ECB_No;
    RootComponent = CollisionComponent;

    ProjectileMovementComponent =
      CreateDefaultSubobject<UProjectileMovementComponent>(TEXT
        ("Projectile"));
    ProjectileMovementComponent->UpdatedComponent =
      CollisionComponent;
    ProjectileMovementComponent->InitialSpeed = 1800.f;
    ProjectileMovementComponent->MaxSpeed = 1800.f;
    ProjectileMovementComponent->bRotationFollowsVelocity = true;
    ProjectileMovementComponent->ProjectileGravityScale = 0.f;

    MeshComponent =
      CreateDefaultSubobject<UStaticMeshComponent>(TEXT("Mesh"));
    MeshComponent->SetupAttachment(RootComponent);
    MeshComponent->SetRelativeRotation(FRotator(0.f, -90.f, 0.f));
    MeshComponent->SetRelativeScale3D(FVector(2.f, 2.f, 2.f));

  static ConstructorHelpers::FObjectFinder<UStaticMesh>
    StaticMeshAsset(
      TEXT("/Game/KayKit/PrototypeBits/Models/Bullet.Bullet"));

  if (StaticMeshAsset.Succeeded())
  {
      MeshComponent->SetStaticMesh(StaticMeshAsset.Object);
  }
}
```

Most of the aforementioned code has been addressed previously or is self-explanatory, although there are some important things to mention. The `CollisionComponent` collision profile name has been set to `BlockAll`, in order to get proper collisions; additionally, we have bound the `OnComponentHit` delegate to the `OnHit()` method, in order to react to any bullet collisions.

We can now add the final method implementation, which will handle the bullet hitting any object:

```
void ABaseGunBullet::OnHit(UPrimitiveComponent* HitComp, AActor*
  OtherActor, UPrimitiveComponent* OtherComp, FVector NormalImpulse,
    const FHitResult& Hit)
{
    if (OtherActor != nullptr && OtherActor != this)
    {
        const auto DamageEvt = FDamageEvent();
        OtherActor->TakeDamage(1.f, DamageEvt, nullptr, nullptr);
    }
    Destroy();
}
```

As you can see, we just call the `TakeDamage()` method to the `Actor` object that has been hit and then we destroy the bullet. No need to worry about damage parameters in this game! They aren't the focus of this book, so you have the freedom to add your own damage logic and stick to it if you'd like. Feel free to customize the game to your heart's content!

Now that we've finalized the bullet class, it's time to create a suitable target for it.

Implementing a Target class

We now need to create a base actor that we will be using to implement a target for our shooting AI agent. So, let's start by creating a new C++ class extending from `Actor` and call it `BaseTarget`. Once the class has been created, open up the `BaseTarget.h` header file.

As a first step, add this forward declaration just after the `#include` section:

```
class UStaticMeshComponent;
```

Next, remove the `BeginPlay()` and `Tick()` functions as they won't be needed, and add the following declarations just after the `ABaseTarget()` constructor:

```
protected:
    UPROPERTY(VisibleAnywhere, BlueprintReadOnly,
      Category="Projectile", meta=(AllowPrivateAccess="true"))
    UStaticMeshComponent* MeshComponent;

    virtual float TakeDamage(float DamageAmount, FDamageEvent const&
      DamageEvent, AController* EventInstigator, AActor* DamageCauser)
        override;
```

Apart from the `StaticMeshComponent` property that will be used to show the target mesh, we have added the `TakeDamage()` declaration that will be used to handle hits from bullets.

Now, open the `BaseTarget.cpp` file and, after removing the `BeginPlay()` and `Tick()` function implementations, change the constructor with the following code:

```
ABaseTarget::ABaseTarget()
{
    PrimaryActorTick.bCanEverTick = false;

    MeshComponent =
        CreateDefaultSubobject<UStaticMeshComponent>(TEXT("Mesh"));
    MeshComponent->SetupAttachment(RootComponent);
    MeshComponent->SetRelativeRotation(FRotator(0.f, -90.f, 0.f));
    RootComponent = MeshComponent;

    static ConstructorHelpers::FObjectFinder<UStaticMesh>
        StaticMeshAsset(
    TEXT("/Game/KayKit/PrototypeBits/Models/target_stand_B_target_
        stand_B.target_stand_B_target_stand_B"));
    if (StaticMeshAsset.Succeeded())
    {
        MeshComponent->SetStaticMesh(StaticMeshAsset.Object);
    }
}
```

By now, you should already be familiar with the previous code; after disabling the tick functionality for this actor, we proceeded to add and initialize a `StaticMesh` component, which will serve the purpose of displaying a target in our project.

Now, add the `TakeDamage()` implementation to your file:

```
float ABaseTarget::TakeDamage(float DamageAmount, FDamageEvent const&
    DamageEvent, AController* EventInstigator,
    AActor* DamageCauser)
{
    Tags[0] = "Untagged";
    return DamageAmount;
}
```

As you can see, the approach here is quite simple; we are merely using the `Untagged` keyword, which will invalidate the tag assigned to the parent object. This effectively renders it invisible to the behavior tree task we will be constructing later in this chapter. We don't need to worry about damage logic; once a target has been it, it will just be invalidated.

Now that all the base classes have been created, we are ready to implement the needed Blueprints, including a brand-new gunner character.

Creating the Blueprints

We will proceed with the creation of the Blueprints that will be instrumental in generating the new AI gunner agent. Specifically, we will be working on the following components:

- The spawnable bullet
- The target
- The gunner character itself

Let's start by creating the bullet Blueprint.

Implementing the bullet Blueprint

The creation of the bullet Blueprint is quite straightforward. Just follow these steps:

1. Open the `Blueprints` folder in the Unreal Engine **Content Drawer**.
2. Right-click on it and select **Blueprint Class**.
3. From the **All Classes** section, select **BaseGunBullet**.
4. Name the new asset `BP_GunBullet`. *Figure 9.2* shows the final Blueprint class:

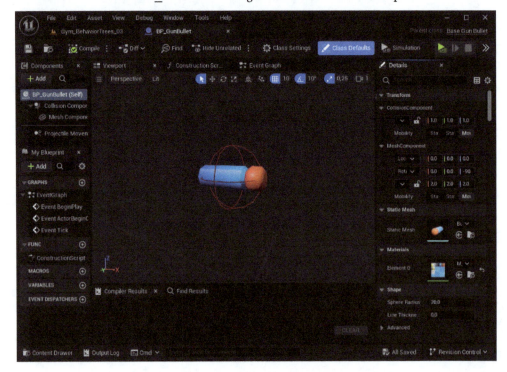

Figure 9.2 – Gun bullet Blueprint

Implementing the target Blueprint

The creation of the target Blueprint is almost identical to the bullet one; we just need to add a tag to the object. Do the following steps:

1. Open the `Blueprints` folder in the Unreal Engine **Content Drawer**.

2. Right-click on it and select **Blueprint Class**.

3. From the **All Classes** section, select **BaseTarget**.

4. Name the new asset `BP_Target`.

5. In the **Details** panel, look for the **Tags** property in the **Actor | Advanced** category and hit the + button to create a new tag.

6. Name the tag `ShootingTarget`, as shown in *Figure 9.3*:

Figure 9.3 – Target tag

Figure 9.4 shows the final Blueprint class:

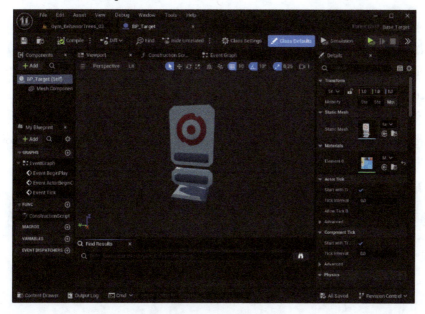

Figure 9.4 – Target Blueprint

Implementing the gunner character Blueprint

It's now time to create the gunner character as a Blueprint. You are already familiar with the process, but here are the steps to follow:

1. Open the `Blueprints` folder in the Unreal Engine **Content Drawer**.

2. Right-click on it and select **Blueprint Class**.

3. From the **All Classes** section, select **BaseDummyCharacter**.

4. Name the new asset `BP_GunnerDummy`.

We'll add the AI controller later in this chapter, but we need to change a value in the Blueprint class in order to make our character properly work. So, open this Blueprint and, in the **Details** panel, locate the **Use Controller Rotation Yaw** attribute in the **Pawn** category; this will allow rotation of the character correctly when using AI logic later on. Tick the value, as depicted in *Figure 9.5*:

Figure 9.5 – Use Controller Rotation Yaw ticked

This value will allow us to rotate the character through a task when shooting at a target.

Now, it's time to add a proper weapon to the character. To do this, follow these steps:

1. With the Blueprint character class open, locate the **Components** panel and click the + **Add** button.

2. Select **UnrealAgilityArena | Base Weapon** to add this component to the Blueprint class.

3. Select this newly added component and, in the **Bullet** category of the **Details** panel, locate the **Bullet Class** attribute; from its dropdown, select **BP_GunBullet**.

Now that we've got our character all primed and ready, it's time to unleash its shooting prowess at our command!

Making the weapon shoot

To make the shooting phase work properly, we will be using **Anim Notify** – an event that can be synchronized with animation sequences – so that we can call the `Shoot()` function at a specific point along the animation timeline.

> **Note**
>
> You might be curious about why we're implementing this particular system instead of directly calling the `Shoot()` function from any part of our code. Well, here's the thing: the shoot animation has a duration, and the moment when the bullet should be spawned occurs somewhere within the animation. This is where an Anim Notify comes into play. By using **Anim Notify**, we can specify the exact moment within the animation when the bullet should be spawned.

An Anim Notify is created by extending the `AnimNotify` class, so start by creating a new C++ class that extends `AnimNotify` and call it `AnimNotify_Shoot`. Once the files have been created, open the `AnimNotify_Shoot.h` header file and, as a first step, change the `UCLASS()` macro declaration to the following:

```
UCLASS(const, hidecategories=Object, collapsecategories, Config =
Game, meta=(DisplayName="Shoot"))
```

Without delving too deeply into the specifics, it is sufficient to say that these initialization settings are necessary for the class to function correctly.

After that, add the following `public` declarations to the class:

```
public:
    UAnimNotify_Shoot();

    virtual void Notify(USkeletalMeshComponent* MeshComp,
        UAnimSequenceBase* Animation, const FAnimNotifyEventReference&
            EventReference) override;
```

The `UAnimNotify_Shoot()` declaration is the constructor, which is quite self-explanatory, while the `Notify()` declaration will be called when an Anim Notify is triggered.

Now, open the `AnimNotify_Shoot.cpp` file and add the needed `#include` declarations at its top:

```
#include "BaseDummyCharacter.h"
#include "BaseWeaponComponent.h"
```

After that, add the constructor implementation:

```
UAnimNotify_Shoot::UAnimNotify_Shoot():Super()
{
#if WITH_EDITORONLY_DATA
    NotifyColor = FColor(222, 142, 142, 255);
#endif
}
```

While not mandatory, this function allows you to customize the color of the notify label within the Unreal Engine Editor. It's quite handy, isn't it?

On the other hand, the Notify() function holds significant importance for gameplay-related reasons, so add the following lines of code:

```
void UAnimNotify_Shoot::Notify(USkeletalMeshComponent* MeshComp,
UAnimSequenceBase* Animation,
                               const FAnimNotifyEventReference&
EventReference)
{
    if(MeshComp == nullptr) return;
    const auto Character = Cast<ABaseDummyCharacter>
      (MeshComp->GetOwner());
    if(Character == nullptr) return;
    const auto WeaponComponent = Character->
      GetComponentByClass<UBaseWeaponComponent>();
    if(WeaponComponent == nullptr) return;
    WeaponComponent->Shoot();
}
```

This function looks for a BaseWeaponComponent instance, if any, and calls the Shoot() function.

Before adding this Anim Notify to the shoot animation, you will need to compile your project. Once the new class is available, look for the **AM_1H_Shoot** montage, which can be found in the Content/ KayKit/PrototypeBits/Character/Animations folder.

> **Note**
>
> In Unreal Engine, a **montage** refers to a specialized asset that allows you to create complex animations for characters or objects. Montages are commonly used for defining sequences of related animation. As montages are not part of this book's focus, I have provided the needed ones for you.

Once you have opened the asset by double-clicking on it, you will notice that, in the asset timeline, there is a **ShootNotify_C** label; this is an empty placeholder I have provided for you to let you know where the notify should be placed.

Figure 9.6 – Animation montage

Right-click on that label and select **Replace** with **Notify | Shoot** to add an `AnimNotify_Shoot` instance.

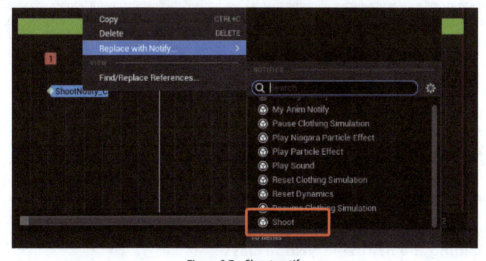

Figure 9.7 – Shoot notify

Now, whenever your AI agent plays this montage, it will get a notification from the montage itself that will call the Shoot() function. Later in this chapter, we will be creating a dedicated task for our AI agent behavior tree, in order to play the montage but, if you want to test things out, you can simply use the **Play Montage** node in the Event Graph of **BP_GunnerDummyCharacter**, as shown in *Figure 9.8*:

Figure 9.8 – Montage test

Just ensure to remove this node after completing your testing to prevent your AI agent from exhibiting seemingly erratic behaviors.

In this section, I have provided insights on enhancing AI agents and behavior trees through best practices and tips from my own experience. Following that, we established the foundation for a more advanced AI agent capable of shooting. The upcoming section will focus on developing a custom decorator, adding a new layer of complexity to our project. Exciting developments lie ahead for our endeavor!

Understanding decorators

Decorators provide a way to add additional functionality or conditions to the execution of a portion of a behavior tree. As you already know from previous chapters, decorators are attached to either a composite or a task node and determine whether a branch in the tree (or even a single node) can be executed. By combining decorators with composite nodes, you can create behavior trees with prioritized behavior allowing for powerful logic capable of handling intricate scenarios. In *Chapter 8*, *Setting Up a Behavior Tree*, we used some built-in decorators but, in this section, I will give you more detailed information about creating your own custom decorators.

Explaining the BTAuxiliaryNode class

Both decorators and services inherit from the BTAuxiliaryNode class, which will let you implement the following functions:

- OnBecomeRelevant(): This will be called when the auxiliary node – the one the decorator or service is attached to – becomes active

- `OnCeaseRelevant()`: This will be executed when the auxiliary node becomes inactive
- `TickNode()`: This will be executed at each auxiliary node tick

In *Chapter 8*, *Setting Up a Behavior Tree*, I presented you with some of these functions, so it's good to know where they come from.

Creating C++ decorators

A decorator extends from the `BTDecorator` class, and in C++, its main implementable functions are as follows:

- `OnNodeActivation()`: This is called when the underlying node is activated
- `OnNodeDeactivation()`: This is called when the underlying node is deactivated
- `OnNodeProcessed()`: This is called when the underlying node is deactivated or fails to activate
- `CalculateRawConditionalValue()`: This computes the value of the decorator condition without considering the inverse condition

Additionally, you can use the `IsInversed()` function to check whether the decorator will handle the inversed conditional value.

Creating Blueprint decorators

Whenever creating a decorator with Blueprints Visual Scripting, you should extend from the `BTDecorator_BlueprintBase` class, which includes some additional code logic and events, in order to better manage it. You can create a decorator in the usual way – from **Content Drawer** – or you can select the **New Decorator** button from the behavior tree graph, as shown in *Figure 9.9*:

Figure 9.9 – Decorator creation

The main events you will have at your disposal when working with Blueprint-generated decorators are as follows:

- **Receive Execution Start AI**: This is called when the underlying node is activated
- **Receive Execution Finish AI**: This is called when the underlying node has finished executing its logic

- **Receive Tick AI**: This is called on each tick

Figure 9.10 – Decorator nodes

By keeping this in mind, you will have the ability to implement your own Blueprint decorators for your AI agents.

We are now going to implement our own decorator, one that will be checking a tag on an actor.

Implementing the CheckTagOnActor decorator

Now is the perfect time to create our first decorator. As you may recall, while implementing the `BaseTarget` class, we ensured that whenever a target gets hit, its tag is set to an undefined value. By implementing a decorator that checks an actor instance tag, we can determine whether the actor itself is a viable target.

So, let's start by creating a new C++ class extending `BTDecorator`, and let's call it `BTDecorator_CheckTagOnActor`. Once the class has been created, open the `BTDecorator_CheckTagOnActor.h` file and add the following declarations:

```
protected:
    UBTDecorator_CheckTagOnActor();

    UPROPERTY(EditAnywhere, Category=TagCheck)
    FBlackboardKeySelector ActorToCheck;

    UPROPERTY(EditAnywhere, Category=TagCheck)
    FName TagName;

    virtual bool CalculateRawConditionValue(UBehaviorTreeComponent&
OwnerComp, uint8* NodeMemory) const override;
```

```
virtual void InitializeFromAsset(UBehaviorTree& Asset) override;
```

As you can see, we will be using a Blackboard key value – the `ActorToCheck` one – to check whether its referred value has a tag equal to `TagName`. This check will be handled by the `CalculateRawConditionValue()` function. Additionally, we will need to initialize any asset-related data, and this is usually done in the `InitializeFromAsset()` function, which is inherited by the `BTNode` superclass.

Now, open the `BTDecorator_CheckTagOnActor.cpp` file to start implementing the functions. Let's start by adding the needed `#include` files:

```
#include "BehaviorTree/BlackboardComponent.h"
#include "BehaviorTree/Blackboard/BlackboardKeyType_Object.h"
```

Next, let's implement the constructor:

```
UBTDecorator_CheckTagOnActor::UBTDecorator_CheckTagOnActor()
{
    NodeName = "Tag Condition";
    ActorToCheck.AddObjectFilter(this, GET_MEMBER_NAME_
      CHECKED(UBTDecorator_CheckTagOnActor, ActorToCheck),
        AActor::StaticClass());
    ActorToCheck.SelectedKeyName = FBlackboard::KeySelf;
}
```

What we are doing here, immediately after naming the node, holds significant importance. We are filtering key values to only allow `Actor` classes. This step ensures that only valid Blackboard keys related to actors will be accepted, maintaining the integrity and appropriateness of the inputs.

The `CalculateRawConditionValue()` function is going to be pretty straightforward:

```
bool UBTDecorator_
CheckTagOnActor::CalculateRawConditionValue(UBehaviorTreeComponent&
OwnerComp,
                                                             uint8*
NodeMemory) const
{
    const UBlackboardComponent* BlackboardComp = OwnerComp.
      GetBlackboardComponent();
    if (BlackboardComp == nullptr) return false;

    const AActor* Actor = Cast<AActor>(BlackboardComp-
      >GetValue<UBlackboardKeyType_Object>(ActorToCheck.
        SelectedKeyName));

    return Actor != nullptr && Actor->ActorHasTag(TagName);
}
```

As you can see, we retrieve the Blackboard component and get the `ActorToCheck` key in order to check whether there is a valid `Actor` instance and whether it is tagged as a target.

Now, implement the last required function:

```
void UBTDecorator_CheckTagOnActor::InitializeFromAsset(UBehaviorTree&
Asset)
{
    Super::InitializeFromAsset(Asset);

    if (const UBlackboardData* BBAsset = GetBlackboardAsset();
      ensure(BBAsset))
    {
        ActorToCheck.ResolveSelectedKey(*BBAsset);
    }
}
```

This function retrieves the `BlackboardData` asset and resolves the selected key for `ActorToCheck` from that asset.

In this section, you have been provided with more advanced information about decorators, including specific considerations for implementing them in C++ or Blueprints. Additionally, you have successfully created a custom decorator that will be utilized by our upcoming gunner AI agent. This custom decorator will play a crucial role in creating the behavior and decision-making capabilities of the AI gunner agent, further improving its performance and effectiveness.

In the next section, I will be presenting you with some detailed information on how to implement services.

Understanding services

Since you are already acquainted with services from previous chapters, I will now provide you with additional information to further enrich your understanding of this topic. Let's explore these details to enhance your expertise in services within behavior trees.

Creating C++ services

A service extends from the `BTService` class, and its main implementable function is `OnSearchStart()`, which is executed when the behavior tree search enters the underlying branch. You can use this to create some kind of initialization if needed.

Furthermore, it is important to remember that services extend the `BTAuxiliaryNode` class, thereby inheriting all of its functions. One particularly crucial function inherited by services is the `TickNode()` function, which plays a vital role in the implementation of services, as it governs the execution and periodic updating of the service node within the behavior tree.

Creating Blueprint services

When creating a service using Blueprints Visual Scripting, it is advisable to extend from the `BTService_` `BlueprintBase` class as it provides additional code logic and events that facilitate better management of the service itself. Similar to decorators, there are two ways to create a service: the conventional method, which involves using **Content Drawer**, or selecting the **New Service** button directly from the behavior tree graph, as shown in *Figure 9.11*:

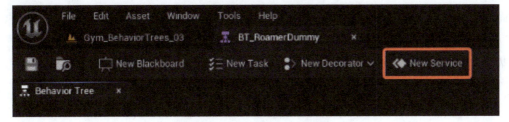

Figure 9.11 – Service creation

The main events you will have at your disposal when working with Blueprint-generated services are as follows:

- **Receive Activation AI**: This is called when the service becomes active

- **Receive Deactivation AI**: This is called when the service becomes inactive

- **Receive Search Start AI**: This is called when the behavior tree search enters the underlying branch

- **Receive Tick AI**: This is called on each tick

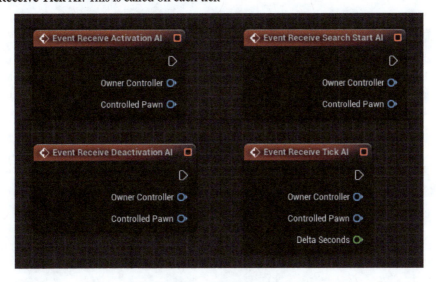

Figure 9.12 – Service nodes

With this in mind, you will have the ability to implement your own Blueprint services for your behavior trees.

With this new knowledge at our disposal, let's implement a new service that will let us handle the ammo state.

Implementing the SetAmmo service

We are now ready to start implementing our own service; you already have created a couple of them in *Chapter 8, Setting Up a Behavior Tree*, so you should be already familiar with some of the presented steps.

In this case, we will need a service that will allow us to tell the Blackboard when a weapon has fired – and so needs reloading – or is ready to shoot. As usual, let's start by creating a new C++ class extending from `BT_Service` and call it `BTService_SetAmmo`. Once it has been created, open the `BTService_SetAmmo.h` file and add the following declarations:

```
public:
    UBTService_SetAmmo();

protected:
    UPROPERTY(BlueprintReadOnly, EditAnywhere, Category="Blackboard")
    FBlackboardKeySelector NeedsReload;

    UPROPERTY(BlueprintReadOnly, EditAnywhere, Category="Blackboard")
    bool bKeyValue = false;

    virtual void OnBecomeRelevant(UBehaviorTreeComponent& OwnerComp,
        uint8* NodeMemory) override;
```

You should already be familiar with most of the code here; let's just say that we will be using a `NeedsReload` Blackboard key as a `bool` value to see whether the weapon ammo is depleted or not. Now, open the `BTService_SetAmmo.cpp` file and add the following `#include` declaration at the top of it:

```
#include "BehaviorTree/BlackboardComponent.h"
```

The constructor is going to be pretty straightforward as we want the service tick to be disabled and we want to execute it just when it becomes relevant:

```
UBTService_SetAmmo::UBTService_SetAmmo()
{
    NodeName = "SetAmmo";
    bCreateNodeInstance = true;
    bNotifyBecomeRelevant = true;
    bNotifyTick = false;
}
```

The `OnBecomeRelevant()` function will serve us just to set the Blackboard key value:

```
void UBTService_SetAmmo::OnBecomeRelevant(UBehaviorTreeComponent&
  OwnerComp, uint8* NodeMemory)
{
    const auto BlackboardComp = OwnerComp.GetBlackboardComponent();
    if (BlackboardComp == nullptr) return;

    BlackboardComp->SetValueAsBool(NeedsReload.SelectedKeyName,
      bKeyValue);
}
```

In this section, you have been provided with additional information about services, including specific considerations for implementing them in C++ or Blueprints. Additionally, you have successfully created another custom service that will be used to handle your AI agent's gun ammo.

In the next section, I will be presenting you with detailed information on how to implement tasks as we will be creating a couple more for our soon-to-be gunner agent.

Understanding tasks

In this section, I will provide you with additional information to enhance your understanding of tasks. Let's explore these details together to further strengthen your grasp of tasks within the context of behavior trees.

Creating C++ tasks

A task extends from the `BTTask` class, and its main implementable functions are as follows:

- `ExecuteTask()`: This will start the task execution and will return the task result
- `AbortTask()`: This will let you handle events where a task should be stopped

This is usually all you need to create even a simple yet fully working task.

Creating Blueprint tasks

When creating a task using Blueprints Visual Scripting, you will be extending from the `BTTask_BlueprintBase` class as it provides additional code logic to facilitate its implementation. As you may have guessed, there are two ways to create a task: the usual creation from **Content Drawer**, and the **New Task** selection button directly from the behavior tree graph, as shown in *Figure 9.13*:

Figure 9.13 – Task creation

The main events you will have available when working with Blueprint-generated tasks are as follows:

- **Receive Execute AI**: This is called when the task is executed
- **Receive Abort AI**: This is called when the task is aborted
- **Receive Tick AI**: This is called on each tick

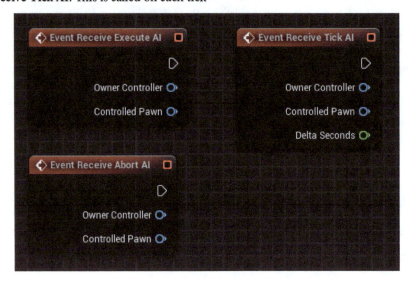

Figure 9.14 – Task nodes

Keeping this in mind, you will possess the capability to implement your own Blueprint tasks for your behavior trees, and that's exactly what we are going to do in the next steps.

Implementing the PlayMontage task

As you already know, the shooting command for our AI agent is going to be controlled by an Anim Notify from an animation montage. Unfortunately, there is no out-of-the-box task for executing montages from a behavior tree; a **PlayAnimation** task is present, but it will not serve our purposes as it won't handle montages really well. Fortunately, with our existing strong understanding of tasks, implementing the code logic will be relatively straightforward. Furthermore, having a task that plays montages will prove highly beneficial when handling other animation sequences, such as reloading bullets or celebrating when all targets have been successfully hit. So, start by creating a new C++ class that extends BTTask and call it BTTask_PlayMontage. Inside the BTTask_PlayMontage.h file, add the following self-explanatory declarations:

```
public:
    UPROPERTY(EditAnywhere, Category="Dummy Task")
    UAnimMontage* AnimMontage;

    virtual EBTNodeResult::Type ExecuteTask(UBehaviorTreeComponent&
        OwnerComp, uint8* NodeMemory) override;
```

Inside the BTTask_PlayMontage.cpp file, add the following implementation:

```
EBTNodeResult::Type UBTTask_
PlayMontage::ExecuteTask(UBehaviorTreeComponent& OwnerComp, uint8*
NodeMemory)
{
    if(AnimMontage == nullptr) return EBTNodeResult::Failed;

    const auto Controller = OwnerComp.GetAIOwner();
    if(Controller == nullptr) return EBTNodeResult::Failed;

    const auto Character = Cast<ACharacter>
      (Controller->GetCharacter());
    if(Character == nullptr) return EBTNodeResult::Failed;
    Character->PlayAnimMontage(AnimMontage, 1.f, FName("Default"));
    return EBTNodeResult::Succeeded;
}
```

This function simply executes the PlayAnimMontage() function on the character, returning a Succeeded result. If any of the needed references are not found, a Failed result is returned.

With this task ready to go, we can implement a second one, the last one we will be needing in this chapter.

Implementing the FindAvailableTarget task

This task will have the sole aim of finding an available target by checking all actors with a predefined tag. There is nothing fancy here, but we will be needing it, so create a new C++ class inheriting from BBTask and call it BTTask_FindAvailableTarget. In the BTTask_FindAvailableTarget.h header file, add the following declarations:

```
public:
    UBTTask_FindAvailableTarget();
```

```
    UPROPERTY(EditAnywhere, Category="Blackboard")
    FBlackboardKeySelector TargetActor;

    UPROPERTY(EditAnywhere, Category="Dummy Task")
    FName TargetTag;

protected:
    virtual EBTNodeResult::Type ExecuteTask(UBehaviorTreeComponent&
        OwnerComp, uint8* NodeMemory) override;
```

No need to add explanations here, so let's open the BTTask_FindAvailableTarget.cpp file
and add the needed #include declarations:

```
#include "BehaviorTree/BlackboardComponent.h"
#include "Kismet/GameplayStatics.h"
```

The constructor is just going to filter the type entries for the TargetValue key:

```
UBTTask_FindAvailableTarget::UBTTask_FindAvailableTarget()
{
    NodeName = "Find Available Target";

    TargetActor.AddObjectFilter(this, GET_MEMBER_NAME_CHECKED(UBTTask_
        FindAvailableTarget, TargetActor), AActor::StaticClass());
    TargetActor.SelectedKeyName = FBlackboard::KeySelf;
}
```

The ExecuteTask() function will search through the level in order to find all Actor instances
correctly tagged and return a random element from the list. Just add this piece of code:

```
EBTNodeResult::Type UBTTask_FindAvailableTarget::ExecuteTask
    (UBehaviorTreeComponent& OwnerComp, uint8* NodeMemory)
{
    const auto BlackboardComp = OwnerComp.GetBlackboardComponent();
    if (BlackboardComp == nullptr) { return EBTNodeResult::Failed; }

    TArray<AActor*> TargetList;
    UGameplayStatics::GetAllActorsWithTag(GetWorld(), TargetTag,
        TargetList);
    if(TargetList.Num() == 0) { return EBTNodeResult::Failed; }

    const auto RandomTarget = TargetList[FMath::RandRange(0,
        TargetList.Num() - 1)];
```

```
BlackboardComp->SetValueAsObject(TargetActor.SelectedKeyName,
    RandomTarget);

return EBTNodeResult::Succeeded;
}
```

As you can see, a `Succeeded` result is returned if at least one `Actor` instance has been found.

In this section, we took a brief look at some of the key features of tasks and even added a couple more to our arsenal. It seems like we are now well-prepared to embark on our journey with the gunner AI character. While we are discussing this topic, it's a great opportunity to explain the proper techniques for debugging a behavior tree. So, let's dive in and get started with it!

Debugging behavior trees

Debugging behavior trees with Unreal Engine is essential for ensuring the smooth and efficient functioning of your AI-driven games. By carefully examining and analyzing the behavior tree's execution, you can identify and resolve any issues or glitches that may arise during gameplay. You already have some understanding of how to enable the debugging tools in Unreal Engine. In this section, we will take a deep dive into the debugging feature specifically designed for behavior trees; before starting with the debugging tools, we'll need to create a proper – and moderately complex – behavior tree.

Creating the Blackboard

The Blackboard for the behavior tree is going to be straightforward; we need a couple of keys to keep a reference of the target and a flag to check whether the weapon needs reloading. So, let's start by doing the following:

1. Open **Content Drawer** and create a Blackboard asset in the `Content/AI` folder.

2. Name the asset `BB_GunnerDummy` and open it.

3. Create a new key of the `bool` type and name it `NeedsReload`.

4. Create a new key of the `Object` type and call it `TargetActor`.

You may remember that while creating the `BTTask_FindAvailableTarget` class, we decided to filter this key so that it will accept only the `Actor` type and not a generic `Object` one; this means you will need to set the base class for this key to an `Actor` type. To do this, follow these steps:

1. Select the **TargetActor** key and, in the **Blackboard Details** panel, open the **Key Type** option to show the **Base Class** attribute.

2.   From the **Base Class** dropdown, select **Actor**, as shown in *Figure 9.15*:

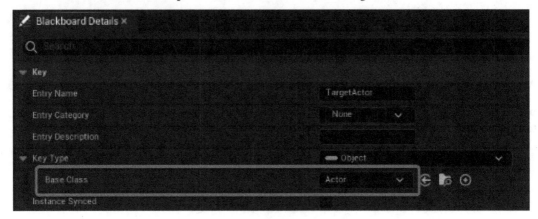

Figure 9.15 – Target Actor key

Creating the behavior tree

The behavior tree we are going to implement is going to have the following logic:

- If the AI character has a valid target, it will shoot at it and then reload the weapon

- If no target is set, it will try to find one in the level and set the proper key

- If no target is available in the level, it means all targets have been hit and the AI character will cheer with happiness

Start by doing the following:

1.   In the Content/AI folder, create a new **Behavior Tree** asset and call it BT_GunnerDummy.

2.   In the **Details** panel, set the **Blackboard Asset** attribute to BB_GunnerDummy.

3. Connect a **Selector** node to the **Root** node and call it `Root Selector`, as shown in *Figure 9.16*:

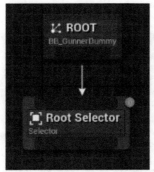

Figure 9.16 – The Root Selector node

As you may remember, a selector node will execute the subtrees in sequence until one of them succeeds; this is exactly what we need to do in order to create our gunner AI logic.

Adding the shooting logic

The shooting logic is going to be subdivided into two phases – shooting and reloading – so we are going to use another selector node. Let's do the following:

1. From **Root Selector**, add another selector node and name it `Shoot Selector`.
2. Right-click on it, add a **CheckTagOnTarget** decorator, and name it `Is Actor a Target?`.
3. Select this decorator and, in the **Details** panel, do the following:

    - Set the **Actor to Check** attribute to **TargetActor**
    - Set the **Tag Name** attribute to **ShootingTarget**

Basically, this selector will be executed only if there is a valid target in the **TargetActor** key of the Blackboard; if not, **Root Selector** will try executing the next subtree available. We need now to create the actual shooting logic, so start doing the following steps:

1. Add a sequence node to **Shoot Selector** and name it `Shoot Sequence`.
2. Right-click on it, add a **Blackboard** decorator, and name it `Has Ammo?`.
3. Select the decorator and, in the **Details** panel, do the following:

    - Set the **Notify Observers** attribute to **On Value Change**
    - Set the **Key Query** attribute to **Is Not Set**
    - Set the **Blackboard Key** attribute to **NeedsReload**

This part of the tree will execute only when the **NeedsReload** key is set to **true**; otherwise, it will try to execute the next subtree. This portion of the tree graph should look like the one depicted in *Figure 9.17*:

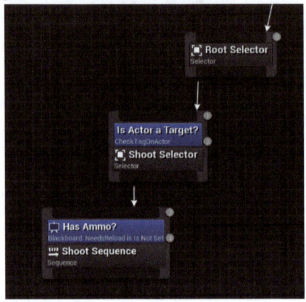

Figure 9.17 – Starting shoot sequence

Let's add some nodes to **Shoot Sequence**:

4. Add a **Rotate to Face BBEntry** task and name it `Rotate Towards Target`.

5. Select this node and, in the **Details** panel, set the **Blackboard Key** attribute to **TargetActor**.

6. From **Shoot Sequence**, add a **Play Montage** task and name it `Shoot Montage`. Make sure that this task is at the right of the **Rotate Towards Target** task. In the **Details** panel, set the **Anim Montage** attribute to **AM_1H_Shoot**.

7. Right-click on this task node, add a **Set Ammo** service, and name it `Deplete Ammo`.

8. Select this service and do the following:

    • Set the **Needs Reload** attribute to **NeedsReload**

    • Check the **Key Value** attribute

9. From **Shoot Sequence**, add a **Wait** node and make sure that this task is at the right of the **Shoot Montage** task. Select the **Wait** node and do the following:

    • Set the **Wait Time** attribute to `2.0`

    • Set the **Random Deviation** attribute to `0.5`

This portion of the behavior tree can be seen in *Figure 9.18*:

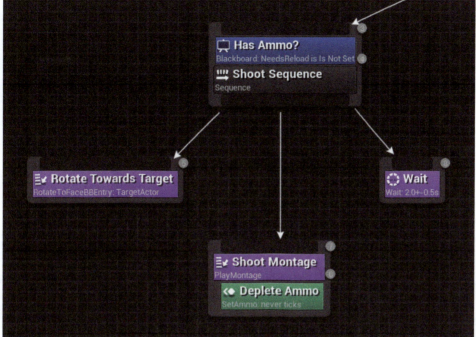

Figure 9.18 – Finished shoot sequence

One of the great things about behavior trees is that when you give descriptive names to your nodes, you can quickly understand what's happening at a glance. By naming your nodes in a way that accurately reflects their purpose or function, you create a clear and intuitive visual representation of your AI.

We can now start creating the reload sequence for the gun. Let's start by following these steps:

1. From **Shoot Selector**, add a new sequence node to the right of **Shoot Sequence** and name it `Reload Sequence`.

2. From **Reload Sequence**, add a **Play Montage** task and name it `Reload Montage`. In the **Details** panel, set the **Anim Montage** attribute to **AM_1H_Reload**.

3. Right-click on this task node, add a **Set Ammo** service, and name it `Refill Ammo`.

4. Select this service and do the following:

   - Set the **Needs Reload** attribute to **NeedsReload**

   - Leave the **Key Value** attribute unchecked

5. From **Reload Sequence**, add a **Wait** node and make sure that this task is at the right of the **Reload Montage** task. Select the **Wait** node and do the following:

- Set the **Wait Time** attribute to 3 . 0.
- Set the **Random Deviation** attribute to 0 . 5. This portion of the behavior tree can be seen in *Figure 9.19*:

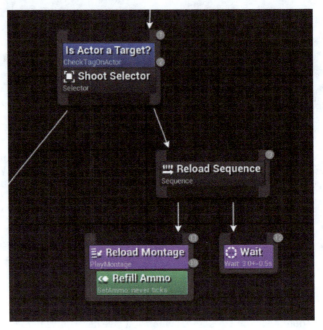

Figure 9.19 – Reload sequence

Whenever we enter this portion of the tree, start the reload animation while setting the **NeedsReload** key to **false**, and we wait a bit before going on with the execution. With this portion of the behavior tree complete, we can implement the target search portion.

Finding an available target

Whenever there is no available target to shoot at, **Root Selector** will execute the next subtree; in this case, we will be looking for a new viable target. To do this, follow these steps:

1. From **Root Selector**, add a **FindAvailableTarget** task at the right of the **Shoot Selector** node.
2. Select the task and do the following:

- Set the **Target Actor** attribute to **TargetActor**
- Set the **Target Tag** attribute to **ShootingTarget**

Figure 9.20 shows this portion of the behavior tree:

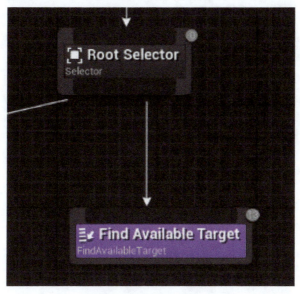

Figure 9.20 – Find target task

It's now time to add the third and last part of the behavior tree logic.

Finishing the AI logic

The last portion of the code will be to make the AI character cheer when all targets have been hit. To do so, follow these steps:

1. From **Root Selector**, add a sequence node at the right of the **Find Available Target** task and name it Cheer Sequence.

2. From **Cheer Sequence**, add a **Play Montage** task and name it Cheer Montage. In the **Details** panel, set the **Anim Montage** attribute to **AM_Cheer**.

3. From **Cheer Sequence**, add a **Wait** node and make sure that this task is at the right of the **Cheer Montage** task. Select the **Wait** node and do the following:

   - Set the **Wait Time** attribute to 3.0

   - Set the **Random Deviation** attribute to 0.5

This part of the graph should look like *Figure 9.21*:

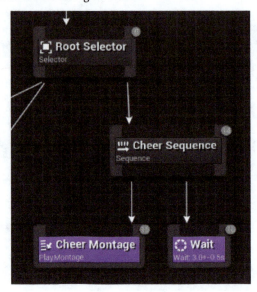

Figure 9.21 – Cheer sequence

Now that the behavior tree is finally complete, we can move forward by creating a dedicated AI controller.

Creating the AI controller

The AI controller is going to be pretty simple; you'll just need to do the following:

1. Open **Content Drawer** and, in the `Content/Blueprints` folder, add a new Blueprint class extending **BaseDummyAIController** and name it `AIGunnerDummyController`.

2. Open it and, in the **Details** panel, locate the **Behavior Tree** property and set its value to **BT_GunnerDummy**.

3. Open **BP_GunnerDummyCharacter** and, in the **Details** panel, set the **AI Controller Class** attribute to **AIDummyGunnerController**.

Now that we have the controller ready and the character all set up, it's time to test and debug its behavior.

Debugging the behavior tree on a gym

To start debugging the newly created behavior tree, let's start by creating a new level. Let's follow these steps:

1. Create a level of your choice, starting from the Level Instances and Packed Level Actors I provided in the project template.

2. Add a **BP_GunnerDummyCharacter** instance.

3. Add one or more **BP_Target** instances so that your AI character will have a line of sight to them. My gym level is shown in *Figure 9.22*:

Figure 9.22 – Gym level

Once you test the level, the expected behavior is for the character to shoot at each target, reload after each shot, and cheer once all targets have been successfully hit.

Adding breakpoints

To test your behavior tree, you can open it up and start the level simulation; you will see the active part of the tree, with nodes highlighted in yellow, as shown in *Figure 9.23*:

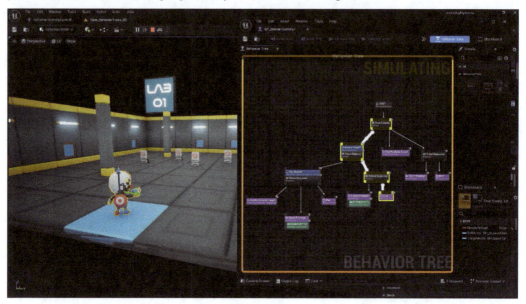

Figure 9.23 – Testing the tree

Sometimes, portions of the graph will be executed really fast and you may not see whether a particular portion of the tree has been executed. To get a better understanding of what's happening, you may add a breakpoint by right-clicking on a node and selecting **Add Breakpoint**, as shown in *Figure 9.24*:

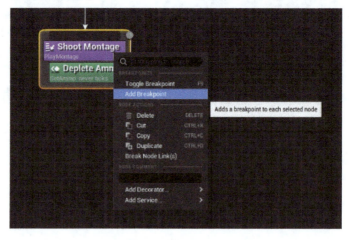

Figure 9.24 – Adding a breakpoint

Note

In Unreal Engine's behavior tree, a **breakpoint** is a debugging feature that allows you to pause the execution of the behavior tree at a specific node. When the execution reaches the breakpoint, the behavior tree execution is temporarily halted, giving you the opportunity to inspect the state of the AI character and analyze the flow of the behavior tree. Execution can be resumed at any time to get on with the behavior tree execution.

When the behavior tree is executed, it will pause at the breakpoint, providing a clear view of what is happening at that moment. By pausing the execution at specific breakpoints, you can gain valuable insights into the inner workings of the AI behavior and identify any issues or unexpected behaviors that need to be addressed. *Figure 9.25* shows a breakpoint positioned on the **Find Available Target** node, showing that the previous subtree failed while checking the **Is Actor a Target?** decorator:

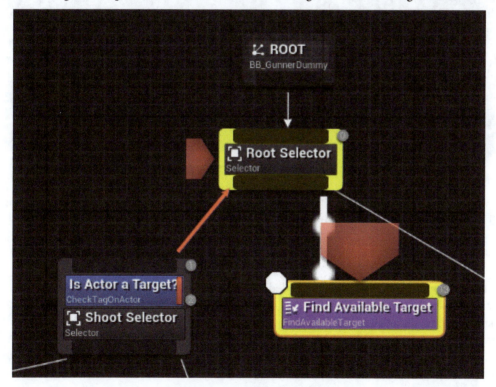

Figure 9.25 – Active breakpoint

Using the debugging tools

As you may recall from *Chapter 6, Optimizing the Navigation System*, Unreal Engine offers a range of debugging tools for the AI system. Behavior trees are no exception, and once you enable these tools, you will have the ability to analyze the situation by pressing the number *2* on your keyboard's numpad. This feature allows you to gain insights into the behavior of the AI character and evaluate the execution of the behavior tree in real time. In *Figure 9.26*, we can observe a specific situation where the behavior tree is displayed on the screen:

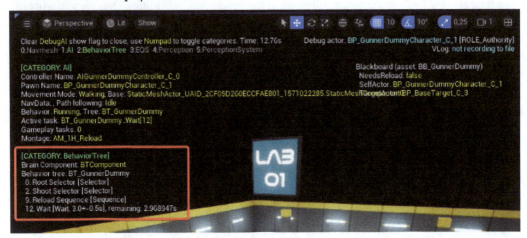

Figure 9.26 – Debugging tools

This visual representation provides a clear view of the structure and flow of the behavior tree, allowing us to analyze and understand its organization.

Summary

In this comprehensive chapter, I have provided you with additional information on creating more effective and efficient behavior trees. We began by discussing some best practices and then delved into the key features of decorators, services, and tasks. Furthermore, we explored the implementation of custom nodes tailored to specific requirements, culminating in the creation of a fully functional AI agent. To validate our work and take advantage of Unreal Engine's robust debugging tools, we also developed a gym environment for thorough testing.

Get ready for the next chapter because things are about to get even more packed with excitement! In this upcoming chapter, I'll be unveiling the Unreal Perception System, where your AI agents will sharpen their senses and become more attentive than ever before!

10

Improving Agents with the Perception System

The AI **Perception System** is a powerful tool in the Unreal Engine Gameplay Framework, as it allows AI-controlled actors to perceive and react to various stimuli in an environment. It provides a way for AI agents to become aware of the presence of other actors – such as players or enemies – through different senses such as sight, hearing, or touch. By properly configuring and using the Perception System, developers can create AI agents that respond appropriately to events in their surroundings. What's more, this system allows developers to implement and configure custom senses tailored to their game's specific needs. This flexibility enables developers to create unique and engaging AI experiences.

In this chapter, we will be covering the main components of the Unreal Engine Perception System, starting with a bit of theory and then applying this newly acquired knowledge to a real-world example.

In this chapter, we will be covering the following topics:

- Presenting the Perception System
- Adding perception to an agent
- Debugging perception
- Creating perception stimuli

Technical requirements

To follow along with the topics presented in this chapter, you should have completed the previous ones and understood their content.

Additionally, if you would prefer to begin with code from the companion repository for this book, you can download the `.zip` project files provided in this book's companion project repository: `https://github.com/PacktPublishing/Artificial-Intelligence-in-Unreal-Engine-5`.

To download the files from the end of the last chapter, click the `Unreal Agility Arena - Chapter 09 - End` link.

Presenting the Perception System

Dr. Markus and Professor Viktoria seem to have a new chapter in their story:

Dr. Markus and Professor Viktoria knew that allowing sophisticated synthetic beings such as their AI dummy puppets to roam unchecked could prove disastrous. Therefore, they started working tirelessly to develop an intricate network of hidden security cameras that could monitor the movements and actions of their creations at all times. With this vigilant surveillance system in place, they hoped to keep them under strict observation and maintain full control, ensuring the safety of their controversial research.

One of the key components when creating intelligent and reactive AI agents in Unreal Engine is the AI Perception System; this powerful system allows AI controllers – and, consequently, AI agents – to perceive and respond to different stimuli in their virtual environment.

At the core of the AI Perception System are **senses** and **stimuli**. A sense – such as sight or hearing – represents a way for an AI agent to perceive its surroundings and is configured to detect specific types of stimuli, which are sources of perception data emanating from other actors in the game world.

As an example, *sight sense* is preconfigured to detect any visible pawn actors, while *damage sense* triggers when the associated AI controller's pawn takes damage from an external source.

> **Note**
> As a developer, you can create custom senses tailored to your game's specific needs by extending the `AISense` class, if you are working with C++, or the `AISense_Blueprint` class, if you are working with Blueprints.

AI Perception System components

The AI Perception System consists of the following main classes:

- `AIPerceptionSystem`: This is the core manager that keeps track of all AI stimuli sources.

- `AIPerceptionComponent`: This represents the AI agent's mind and handles processing perceived stimuli. It needs to be attached to an AI controller to properly work.

- `AIPerceptionStimuliSourceComponent`: This component is added to actors that can generate stimuli and is in charge of broadcasting perception data to listening elements.

- `AIPerceptionSenseConfig`: This defines the properties of a specific sense, what actors can be perceived, and how perception decays over time or distance.

When an actor with `AIPerceptionStimuliSourceComponent` generates a stimulus, nearby `AIPerceptionComponents` detect it through their configured senses. This perceived data is then processed by the AI controller to trigger desired behaviors.

Once you have added `AIPerceptionComponent` to an AI controller, you will need to add one or more `AIPerceptionSenseConfig` elements in order to give your AI agent dedicated senses. *Figure 10.1* shows an example where perception is based on touch:

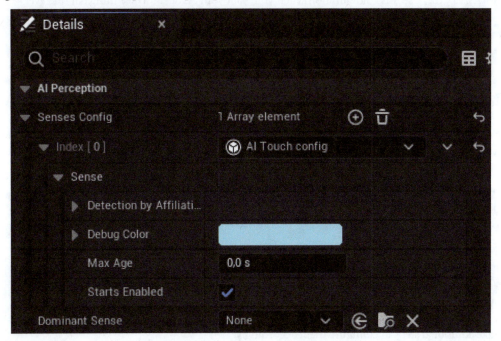

Figure 10.1 – The touch sense config

From the previous screenshot, you may have noticed a **Dominant Sense** property; this property allows you to designate a specific sense that takes priority over others when determining the location of a sensed actor.

Let's explore the available sense configs that, as mentioned earlier, define the properties of each specific sense:

AIPerceptionSenseConfig types

Unreal Engine offers a range of predefined `AIPerceptionSenseConfig` classes that are highly likely to meet your specific requirements. Let's take a look at the available options:

- `AIDamage`: Use this configuration if your AI agent needs to respond to damage events such as *Any Damage*, *Point Damage*, or *Radial Damage*

- `AIHearing`: Use this configuration if you need to detect sounds generated in the surrounding environment

- `AIPrediction`: Use this configuration when you need to predict the target actor location in the next few moments

- `AISight`: Use this configuration when you want your AI agent to see things in the level

- `AITeam`: Use this configuration if you want to notify the AI agent that some ally is nearby

- `AITouch`: Use this configuration when the AI agent touches some other actor or, vice versa, when something is touching the AI agent

Stimuli source

The `AIPerceptionStimuliSourceComponent` class allows an actor to register itself as a source of stimuli for one or more senses. For instance, you can register an actor as a stimuli source for sight. This registration allows an AI agent to visually perceive the actor in the game level.

A stimuli source can be registered – or unregistered – for a sense, making it detectable – or undetectable – by the Perception System.

> **Note**
> The `Pawn` and `Character` classes in Unreal Engine are inherently visible to `AISight` perception due to their default behavior as stimuli sources. This design choice streamlines AI behavior development by eliminating the need to manually configure visibility for each character or pawn. However, if you want the AI to ignore specific characters, you'll need to take additional steps to configure them accordingly.

In this section, we were introduced to the Perception System and its main elements. In the next section, we will work on a fully functional AI agent that will let you sense other actors in your game.

Adding perception to an agent

In this section, we will create a new AI agent that will use the Perception System. We will create a security camera that will probe nearby surrounding areas, looking for some possible targets for the dummy gunner that we created in *Chapter 9*, *Extending Behavior Trees*. Think of it as some kind of infrared camera for a dark environment. Once the camera spots a target, it will tag it so that the gunner will be able to locate it in the environment.

We will start by creating an `Actor` class that will be used as the camera model.

Creating the BaseSecurityCam class

Even though we will be implementing the Perception System inside the AI controller, a nice model to display in the level will help your environment's look and feel, so let's start by creating a new C++ class, extending from the `Pawn` class and named `BaseSecurityCam`. Once the class has been created, open the `BaseSecurityCam.h` file and add the following forward declaration after the `#include` declarations:

```
class UAIPerceptionComponent;
```

Then, make the class a `Blueprintable` one by changing the `UCLASS()` macro to the following:

```
UCLASS(Blueprintable)
```

After that, remove the `BeginPlay()` and `Tick()` declarations, as we won't be using them.

As a final step, add the following component declarations for the static meshes that will display the model just after the `GENERATED_BODY()` macro:

```
UPROPERTY(VisibleAnywhere, BlueprintReadOnly, Category="Security Cam",
  meta=(AllowPrivateAccess="true"))
    UStaticMeshComponent* SupportMeshComponent;

UPROPERTY(VisibleAnywhere, BlueprintReadOnly, Category="Security Cam",
  meta=(AllowPrivateAccess="true"))
UStaticMeshComponent* CamMeshComponent;
```

You can now open the `BaseSecurityCam.cpp` file to implement this class; as a first step, remove the `BeginPlay()` and `Tick()` functions. Then, locate the constructor and change this line of code:

```
PrimaryActorTick.bCanEverTick = true;
```

Change it to this:

```
PrimaryActorTick.bCanEverTick = false;
```

Now, inside the constructor and just after the aforementioned line of code, add the following block of code:

```
SupportMeshComponent =
CreateDefaultSubobject<UStaticMeshComponent>(TEXT("Support Mesh"));
RootComponent = SupportMeshComponent;

static ConstructorHelpers::FObjectFinder<UStaticMesh>
SupportStaticMeshAsset(
    TEXT("/Game/_GENERATED/MarcoSecchi/SM_SecurityCam_Base.SM_
SecurityCam_Base"));

if (SupportStaticMeshAsset.Succeeded())
{
    SupportMeshComponent->SetStaticMesh(SupportStaticMeshAsset.
Object);
}

CamMeshComponent =
CreateDefaultSubobject<UStaticMeshComponent>(TEXT("Cam Mesh"));
CamMeshComponent->SetRelativeLocation(FVector(61.f, 0.f, -13.f));
CamMeshComponent->SetupAttachment(RootComponent);

static ConstructorHelpers::FObjectFinder<UStaticMesh>
  CamStaticMeshAsset(
    TEXT("/Game/_GENERATED/MarcoSecchi/SM_SecurityCam.SM_
SecurityCam"));

if (CamStaticMeshAsset.Succeeded())
{
    CamMeshComponent->SetStaticMesh(CamStaticMeshAsset.Object);
}
```

You already know all about this from the previous chapters of the book, so I suppose there's no need to explain it again. With the security camera model created, we can now implement the corresponding AI controller, along with its perception sense.

Creating the BaseSecurityCamAIController class

To add a proper controller for the security camera, let's create a C++ class extending `AIController` and name it `BaseSecurityCamAIController`. Once the class has been created, open the `BaseSecurityCamAIController.h` file and add the following forward declarations, just after the `#include` declarations:

```
struct FAIStimulus;
struct FActorPerceptionUpdateInfo;
class UBehaviorTree;
```

Then, make the class `Blueprintable` by changing the `UCLASS()` macro to the following:

```
UCLASS(Blueprintable)
```

After that, add this block of code just after the already existing constructor declaration:

```
protected:
    UPROPERTY(EditAnywhere, BlueprintReadOnly, Category = "Dummy AI
Controller")
    TObjectPtr<UBehaviorTree> BehaviorTree;

  virtual void OnPossess(APawn* InPawn) override;

  UFUNCTION()
  void OnTargetPerceptionUpdate(AActor* Actor, FAIStimulus
    Stimulus);
```

You are already familiar with the behavior tree property and the `OnPosses()` function from *Chapter 8, Setting Up a Behavior Tree*; in addition, the `OnTargetPerceptionUpdate()` function will be used as an event handler when getting information from the Perception System.

You can now open `BaseSecurityCamAIController.cpp` and add the following `#include` declarations to the top of the file:

```
#include "Perception/AIPerceptionComponent.h"
#include "Perception/AISenseConfig_Sight.h"
```

Now, locate the constructor, and inside of it, add the following block of code:

```
const auto SenseConfig_Sight = CreateDefaultSubobject<UAISenseConfig_
  Sight>("SenseConfig_Sight");
SenseConfig_Sight->SightRadius = 1600.f;
SenseConfig_Sight->LoseSightRadius = 3000.f;
```

```
SenseConfig_Sight->PeripheralVisionAngleDegrees = 45.0f;
SenseConfig_Sight->DetectionByAffiliation.bDetectEnemies = true;
SenseConfig_Sight->DetectionByAffiliation.bDetectNeutrals = true;
SenseConfig_Sight->DetectionByAffiliation.bDetectFriendlies = true;
```

As you can see, we are creating the sense config for the sight perception, along with some of its properties, such as `SightRadius` and `LoseSightRadius`, which will determine the distance at which the Perception System can detect something and the distance at which detection will be lost, respectively. As redundant as these two attributes may seem, keep in mind that, once a target has been detected, it will be more difficult to lose perception of it, unless both attributes have the same value. `PeripheralVisionAngleDegrees` will handle the cone that will be used to check whether an actor is in the line of sight or not. Lastly, the `DetectionByAffiliation` property is used to handle whether the detected actor is an enemy, friend, or neutral; in this case, we want to check all of them in order to detect anything that is in the line of sight.

Now, it's time to add the actual perception component, so add the following piece of code just after the previous one:

```
PerceptionComponent =
CreateDefaultSubobject<UAIPerceptionComponent>(TEXT("Perception"));
PerceptionComponent->ConfigureSense(*SenseConfig_Sight);
PerceptionComponent->SetDominantSense(SenseConfig_Sight-
  >GetSenseImplementation());
PerceptionComponent->OnTargetPerceptionUpdated.AddDynamic(this,
  &ABaseSecurityCamAIController::OnTargetPerceptionUpdate);
```

As you can see, we create the `AIPerceptionComponent` instance, and then we assign the previously created sight configuration. Lastly, we register to the `OnTargetPerceptionUpdated` delegate that will notify the component of any changes detected by the Perception System.

Now, it's time to implement the `OnPosses()` function, something that we already know how to handle:

```
void ABaseSecurityCamAIController::OnPossess(APawn* InPawn)
{
    Super::OnPossess(InPawn);

    if (ensureMsgf(BehaviorTree, TEXT("Behavior Tree is nullptr!
Please assign BehaviorTree in your AI Controller.")))
    {
        RunBehaviorTree(BehaviorTree);
    }
}
```

The last step is to implement the event handler. To do so, add the following block of code:

```
void ABaseSecurityCamAIController::OnTargetPerceptionUpdate(AActor*
Actor, FAIStimulus Stimulus)
{
    if (Actor->Tags.Num() > 0) return;

    const auto SightID = UAISense::GetSenseID<UAISense_Sight>();


    if (Stimulus.Type == SightID && Stimulus.WasSuccessfullySensed())
    {
        Actor->Tags.Init({}, 1);
        Actor->Tags[0] = "ShootingTarget";
    }
}
```

This function first checks whether the target has already been tagged; in this case, it means that it has already been spotted. It then retrieves the ID of the sight sense, calling GetSenseID(), and checks whether the stimulus type is equal to the sight sense ID and whether the stimulus was successfully sensed. If both conditions are true, it initializes the Tags array, with the first element set to a value of ShootingTarget, in order to make it a viable target for the dummy gunner we have at our disposal.

The security camera is now ready to go; we just need a nice environment to test it on.

In this section, we've shown you how to properly create an AI agent, taking advantage of the Perception System. In the next section, we will test this agent and learn how to properly debug perception information.

Debugging perception

It's now time to test our perception logic and learn how to properly debug the Perception system at runtime. In order to do this, we will need to add some small improvements to the base dummy character. As previously mentioned, the Pawn and Character classes are already registered with sight stimuli, so we won't need to implement this logic. However, we will need to handle damage, as we will be playing around with both BP_RoamerDummyCharacter and BP_GunnerDummyCharacter. It appears that exciting and enjoyable times are just around the corner!

Enhancing the roamer behavior tree

The first step in improving our AI agents will be adding some logic to handle damage to the dummy roamer behavior tree. In particular, we want the AI agent to sit down when it is hit by a Nerf gun projectile. We will start by adding a new key to the dedicated Blackboard.

Improving the Blackboard

A new flag is required for the Blackboard to effectively monitor and keep a record of the character that has been hit. So, open up the BB_Dummy asset and do the following:

1.   Click the **New Key** button, and from the dropdown menu, select **Bool**.

2.   Name the newly created key IsHit.

As you already know, this will expose a new key available to the behavior tree; additionally the key will be exposed to the AI controller, as we will see later on.

Improving the behavior tree

The behavior tree needs to manage the AI agent being hit; in our case, we want to play a montage with the character sitting down, as it has been eliminated from the game. So, let's start by opening the BT_RoamerDummy asset and doing the following:

1.   Right-click on the **Root Sequence** node and add a **Blackboard** decorator, naming it Is Not Hit?.

2.   With the decorator selected, do the following:

     •   Set the **Notify Observer** attribute to **On Value Change**

     •   Set the **Observer aborts** attribute to **Self**

     •   Set the **Key Query** attribute to **Is Not Set**

     •   Set the **Blackboard Key** attribute to **IsHit**

3.   Rename the root sequence In Game Sequence and disconnect it from the **ROOT** node.

4.   Add a **Selector** node to the **ROOT** node and name it Root Selector.

5.   Connect the **Root Selector** node to the **In Game Sequence** node.

6.   Connect the **Root Selector** node to a **PlayMontage** task, and name the newly created node Sit Montage. This node should be at the right of the **In Game Sequence** node.

7.   With the **Sit Montage** node selected, set the **Anim Montage** attribute to AM_Sit. The modified portion of the behavior tree is depicted in *Figure 10.2*:

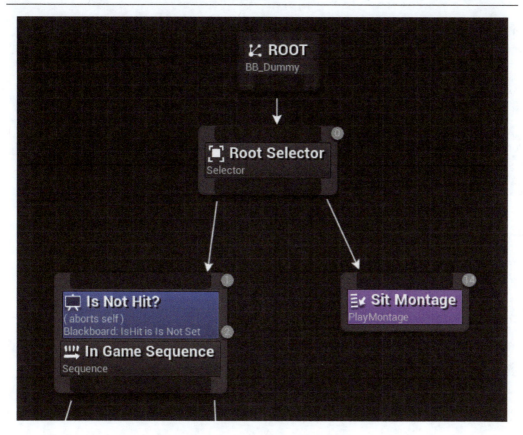

Figure 10.2 – A modified behavior tree

As you can see, the behavior tree will keep on working as before, unless the AI agent has been hit; in that case, the character will sit down and stop wandering.

It's now time to improve the AI controller, in order to manage incoming damage.

Enhancing BaseDummyAIController

The AI controller for any dummy character will need to handle any damage. We are handling all of this inside the AI controller instead of the character for the sake of simplicity; we will need to communicate with the Blackboard, and this is much simpler and more direct when done from the controller itself.

Let's start by opening the `BaseDummyAIController.h` file and adding the following declaration for the damage handler:

```
UFUNCTION()
void OnPawnDamaged(AActor* DamagedActor, float Damage, const
  UDamageType* DamageType, AController* InstigatedBy, AActor*
    DamageCauser);
```

Now, open the `BaseDummyAIController.cpp` file, and in the `OnPossess()` function, add the following line of code:

```
GetPawn()->OnTakeAnyDamage.AddDynamic(this,
&ABaseDummyAIController::OnPawnDamaged);
```

Next, add the following implementation:

```
void ABaseDummyAIController::OnPawnDamaged(AActor* DamagedActor, float
Damage, const UDamageType* DamageType,
    AController* InstigatedBy, AActor* DamageCauser)
{
    const auto BlackboardComp = GetBlackboardComponent();
    BlackboardComp->SetValueAsBool("IsHit", true);

    if (DamagedActor->Tags.Num() > 0)
    {
        DamagedActor->Tags[0] = "Untagged";
    }
}
```

This function is called when a pawn associated with the AI controller is damaged; the function retrieves the Blackboard component of the AI controller and sets the **IsHit** key to a value of `true`. Then, it sets the first tag of the actor to a value of `Untagged` so that it won't be a viable target anymore for the gunner dummy.

With this AI controller all set up, it's time to create a Blueprint for the security camera.

Creating security camera Blueprints

Now, get back to the Unreal Engine Editor, and after compilation has finished, create a Blueprint out of the `BaseSecurityCamAIController` class, naming it `AISecurityCamController`. You don't need to add a behavior tree, as all the logic is handled inside the controller itself.

Now, create a Blueprint class from the `BaseSecurityCam` class, and name it `BP_SecurityCam`. Once it has been created, open it, and in the **Details** panel, locate the **AI Controller Class** attribute and set its value to `AISecurityCamController`.

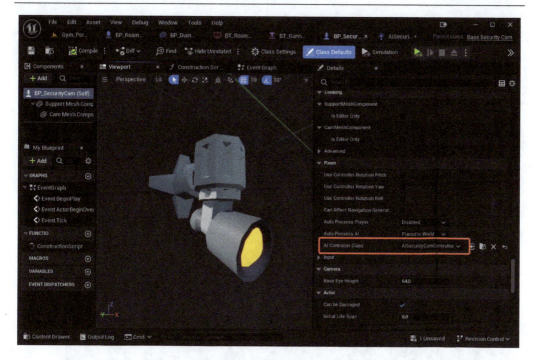

Figure 10.3 – The security cam Blueprint

We have gathered all the necessary elements to bring our new gym to life, and we are ready to take the next steps and start the process of debugging the Perception System.

Creating the gym

We are now going to create a level to test and debug everything. We want to achieve this kind of behavior:

- One or more AI agents will move around the level

- A security cam will try to spot AI agents and tag them as viable targets

- A gunner will wait for the AI agents to be tagged, in order to shoot at them

So, let's start by creating the gym:

1. Create a level of your choice, starting from the Level Instances and Packed Level Actors I provided in the project template.

2. Add a **NavMeshBoundsVolume** actor so that it will cover all the walkable areas.

3. Add some obstacles to make things more interesting.

4. Add a **BP_GunnerDummyCharacter** instance to the level.

5. Add one or more **BP_RoamerDummyCharacter** instances.

6. Add some **NS_Target** Niagara actors that will work as target points for the pathfinding system; just remember to tag them `TargetPoint`. Make sure that the path will bring the AI agents in the line of sight of the gunner.

7. Add one or more **BP_SecurityCam** instances to the walls. The final result should be similar to *Figure 10.4*:

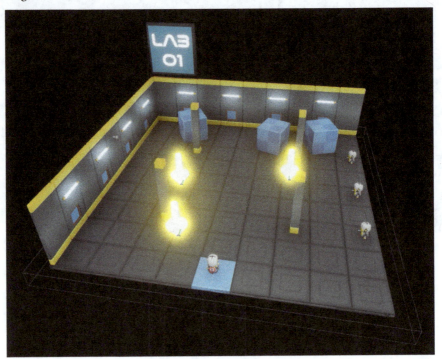

Figure 10.4 – The gym

Considering the type of scenario, where the gunner character will shoot at targets located by the security camera, I have decided to make the gym a bit juicier by adding a post-process volume that simulates an infra-red scenario, as depicted in *Figure 10.5*:

Figure 10.5 – The gym with a post-process volume

This is obviously not mandatory, and you are free to set your post-process environment as you wish.

Now that the gym is finished, it's time to start testing it and learn how to debug the Perception System.

Enabling perception debugging

If you start the simulation, you should see the following things happening:

- As the roamers wander around, the gunner, unaware of their presence, will cheer
- As soon as one of the roamers enters the camera's line of sight, the gunner will start aiming at it
- Every time a roamer is hit, it will sit down and stop wandering

You can tweak the security camera parameters to make it more or less attentive to what's happening in the level.

At this point, you might be curious about how we determine whether an agent is in the camera's line of sight. Well, it's actually quite simple to observe once you enable the debugging tools!

So, let's start by enabling the debugging tools, as explained in *Chapter 6, Optimizing the Navigation Mesh*.

> **Note**
>
> Once the simulation starts, you may be wondering why the security camera has a tiny red icon on top of it, while all the dummy puppets have a green one, as displayed in *Figure 10.6*. When the AI debugging tools are enabled, a green icon is displayed if a pawn has some AI logic set up and running; otherwise, the icon will be red. In our case, the security cam is possessed by a dedicated AI controller, but it has no behavior tree, so the icon will be red.
>
>
>
> Figure 10.6 – AI icons

Now, while the simulation is going on, select a security camera and enable the **Perception** and **Perception System** tools by pressing the *4* and *5* keys on your numpad, respectively. You should immediately see a visualization of the security camera sight sense, as depicted in *Figure 10.7*:

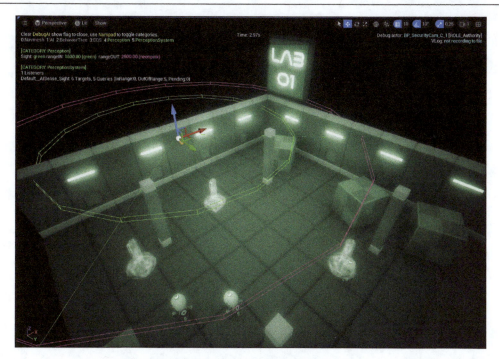

Figure 10.7 – Perception debugging tools

The display will show some important information on the AI agent perception, such as the active sense and its data. Additionally, you will also see a visual representation of the agent sense. In particular, the sight sense will show the following:

- A green circular area that represents the range of the agent sight.

- A pink circular area that represents the maximum range of the agent sight. Once the spotted agent goes beyond this range, sight contact will be lost.

- A green angle that represents the peripheral vision of the agent.

Once an AI agent enters the green circle, it will be detected by the Perception System, and you should see a green line, starting from the security camera and ending at the detected pawn, as depicted in *Figure 10.8*:

Figure 10.8 – A pawn detected

A green wireframe sphere will show the detection point, which will be updated as the detected AI agent moves around.

Once an AI agent moves out of sight, the detection point will stop following it, and you should see the **age** label next to the sphere, updating its value; this is the time that has passed since detection was lost. *Figure 10.9* shows this scenario:

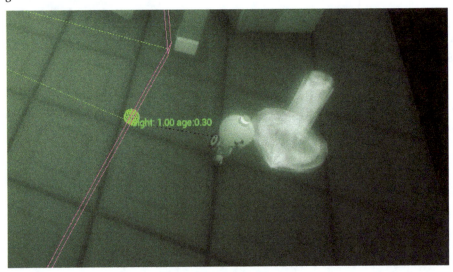

Figure 10.9 – Detection lost

Each sense has its own way of displaying info, so my advice is to start experimenting with each of them to get an understanding of how to debug and get information from the debugging tools.

In this section, we have created a new gym and tested how the Perception System works; what's more, I have shown you how to enable the debugging tools and get a better understanding of what's happening in the level at runtime. In the next section, we are going to take a look at perception stimuli, in order to make our levels more articulated and engaging.

Creating perception stimuli

In the previous section, we utilized the out-of-the-box pawn feature that enables it to be visible to sight sense. We will now analyze actors that are not perceivable by default; this means we will need to add `AIPerceptionStimuliSourceComponent` to an actor. What's more, we will learn how to register or unregister these stimuli, in order to make the actor visible or invisible to the Perception System.

Creating the target actor

In this subsection, we will create a new actor that will serve as a target for the dummy gunner puppet, but with a twist – this actor will create some interference in the level and won't be visible to the security camera. Achieving this kind of feature is quite easy, once you know how to register and unregister an actor from the Perception System. We are basically creating a scrambler device that will disturb – that is, it will be invisible to – the gunner sight sense.

To keep things simple, I will create a Blueprint class; when working with stimuli, it is often more convenient to configure settings directly from a Blueprint rather than using a C++ class. By utilizing a Blueprint, we can easily adjust and fine-tune various aspects of the stimuli, making the whole process more flexible and accessible. This approach allows for quicker iterations and modifications, ultimately resulting in a smoother and more efficient workflow.

To create our scrambler, open the `Blueprints` folder, create a new Blueprint class extending from **Actor**, and name it `BP_Scrambler`. Once the blueprint is opened, follow these steps:

1. In the **Components** panel, add a static mesh component.
2. In the **Details** panel, set the **Static Mesh** property to `SM_RoboGun_BaseRemote` and the **Scale** property to `(3.0, 3.0, 3.0)`.
3. Add an **AIPerceptionStimuliSource** component.
4. In the **Details** panel, locate the **Register as Source for Senses** attribute, add a new element by clicking the + button, and set the value to **AISense_Sight**.

5. Leave the **Auto Register as Source** attribute unchecked

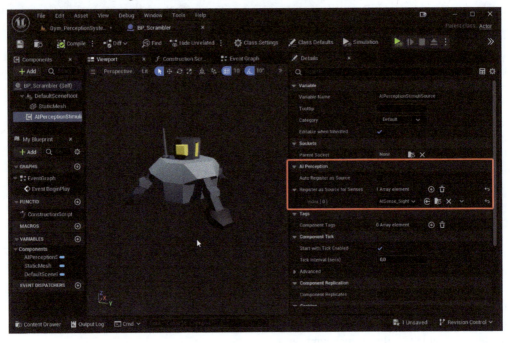

Figure 10.10 – The stimuli source

Now, open the Event Graph for this Blueprint and do the following:

6. From the outgoing execution pin of the **Event Begin Play** node, add a **Delay** node.

7. From the **Completed** pin of the **Delay** node, add a **Register for Sense** node; this should automatically add an **AIPerception Stimuli Source** reference to the **Target** incoming pin.

8. From the **Duration** pin of the **Delay** node, add a **Random Float in Range** node, setting its **Min** and **Max** values to 4.0 and 6.0, respectively.

9. From the dropdown menu of the incoming **Sense Class** pin, select **AISense_Sight**. The final graph should look like the one depicted in *Figure 10.11*:

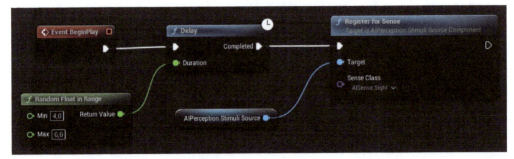

Figure 10.11 – The Event Graph

This graph will simply register the sight sense for this actor after a random interval, making the actor itself visible to the Perception System in the level.

> **Note**
>
> If you need to unregister a stimuli source, the corresponding Blueprint node is **Unregister from Sense**.

Let's test this functionality in a brand-new gym.

Testing the gym

To create the testing level, follow these steps:

1. Create a level of your choice, starting from the Level Instances and Packed Level Actors that I provided in the project template.
2. Add some obstacles to make things more interesting.
3. Add **BP_GunnerDummyCharacter** to the level.
4. Add one **BP_Scrambler** instance so that the gunner puppet can shoot at it.
5. Add one **BP_SecurityCam** instance to the walls so that it is in the line of sight of the scrambler. The final result should be similar to *Figure 10.12*:

Figure 10.12 – The scrambler gym

By starting the simulation, you can see that the scrambler will go unnoticed by the security camera until it reveals itself after a random interval. After that, the scrambler will be tagged as a viable target and the gunner will shoot at it. Try enabling the debugging tools to check what's happening to the Perception System.

In this final section, I have presented how to make any actor detectable by the Perception System. By following the steps and guidelines provided, you can seamlessly integrate the Perception System into your project, allowing your actors to be accurately recognized and interacted with within the virtual environment.

Summary

In this chapter, you learned the basics of the Unreal Engine Perception System. Firstly, we showed the main elements that will let you add senses to your AI agents; after that, you built a pawn with a sight sense that detects moving characters around the level. Then, you learned how to debug the active senses at runtime. Finally, you added a stimuli source to an actor, in order to make it detectable by the Perception System itself. All of this opens up a world of possibilities for creating immersive and dynamic experiences, using the power of the Unreal Engine AI Framework.

In the upcoming chapter, I'll unveil a new method to gather data from the environment; brace yourself, as this cutting-edge feature is still in the experimental stage. But fear not, my friend, for it is knowledge well worth acquiring!

11

Understanding the Environment Query System

The **Environment Query System** (**EQS**) in Unreal Engine is a powerful feature within the AI framework that allows developers to collect data about the virtual environment by letting AI agents query the environment and make informed decisions based on the returned results. In this chapter, you will learn how to properly set up an environment query and how to integrate it inside the behavior tree of an AI agent.

By mastering the EQS, you'll gain the power to create intelligent AI systems that can make informed decisions based on their surroundings. Whether it's finding the best vantage point, locating crucial resources, or strategizing for optimal gameplay, the EQS opens a world of possibilities. While still an experimental feature, learning about the EQS will let you create intelligent and dynamic AI systems in Unreal Engine.

In this chapter, we will be covering the following topics:

- Introducing the Environment Query System
- Setting up an environment query
- Handling environment queries within a behavior tree
- Displaying EQS information

Technical requirements

To follow the topics presented in this chapter, you should have completed the previous ones and understood their content.

Additionally, if you would prefer to begin with code from the companion repository for this book, you can download the `.zip` project files provided in this book's companion project repository: `https://github.com/PacktPublishing/Artificial-Intelligence-in-Unreal-Engine-5`.

To download the files from the end of the last chapter, click the `Unreal Agility Arena – Chapter 10 - End` link.

Introducing the Environment Query System

Well, it seems Dr. Markus is making some progress in his secret lab:

Deep within the hidden confines of their secret laboratory, Dr. Markus and Professor Viktoria toiled tirelessly on their latest endeavor; they were determined to revolutionize their AI dummy puppets by granting them the ability to analyze and probe their environment in unprecedented ways.

With their minds brimming with excitement, Dr. Markus and Professor Viktoria meticulously crafted intricate algorithms as they imbued their creations with an insatiable hunger for knowledge, equipping them with experimental methods to observe and interact with the world around them. As the AI dummy puppets awakened, their eyes flickered with a newfound spark of intelligence, each one exploring its surroundings and analyzing the environment.

Unreal Engine's EQS is a powerful tool that allows developers to define complex queries to gather information about the game world. The EQS enables developers to create AI behaviors that can dynamically adapt to changing environmental conditions. By using the EQS, NPCs or other game entities can make intelligent decisions based on their surroundings. With its flexibility and ease of use, the EQS is a valuable feature in Unreal Engine for creating immersive and interactive gameplay experiences.

With the EQS, you can inquire about gathered data using a set of **tests** that will generate elements that align most closely with the nature of the query posed.

An **EQS query** can be triggered inside a behavior tree – or, alternatively, through scripting – to guide decisions based on the result of tests. These queries mainly consist of **generators** (elements that determine the locations or actors to be tested and weighted) and **contexts** (elements that provide a reference frame for tests or generators). EQS queries empower AI characters to locate optimal positions for tasks such as attacking a player with a line of sight, retrieving health or ammo pickups, or seeking the nearest cover point, among other options.

Let's start by examining all these elements in detail.

Explaining generators

A generator creates the locations or actors – known as **items** – that will undergo testing and weighting; the result will be returned to the behavior tree to which the query belongs. Generators that are available out of the box are as follows:

- **Actors of Class**: This will find all actors of a given class returning them as items for tests
- **Composite**: This will let you create an array of generators and use them for tests
- **Current Location**: This will let you get the location of a specified context and use it to validate tests

- **Points**: Generators can be used to create shape-based traces – **Circle**, **Cone**, **Donut**, **Grid**, and **Pathing Grid** – around a predefined location

What's more, you may implement your own by extending the `EnvQueryGenerator` class (if developing in C++) or `EnvQueryGenerator_BlueprintBase` (if developing with Blueprints).

> **Note**
> Generators created in C++ typically run faster than those developed in Blueprints.

Explaining contexts

A context supplies a point of reference for the different tests and generators and can range from the **querier** – the pawn currently possessed by the AI controller executing the behavior tree – to more complex scenarios involving all actors of a certain type. A generator, such as **Points: Circle**, can use a context that will provide multiple locations or actors.

Available `Context` classes are as follows:

- **EnvQueryContext_Item**: This represents either a location – as a vector – or an actor
- **EnvQueryContext_Querier**: This represents the querier executing the behavior tree

As you may have guessed, you may implement your own context by extending the `EnvQueryContext` class if developing in C++ or `EnvQueryContext_BlueprintBase` if developing with Blueprints.

Explaining tests

A test determines the criteria used by the **environment query** – the actual request to the environment – to select the optimal Item from the generator, given a context.

Some of the out-of-the-box available tests are as follows:

- **Distance**: This will return the distance between the item location and another location
- **Overlap: Box**: This can be used to check whether an item is within the bounds defined by the test itself
- **PathExists: from Querier**: This can be used to check whether a path to the context exists and will return some useful information about it, such as how long the path is

You may implement your own tests by extending the `EnvQueryTest` class both in C++ and Blueprints.

Now that you have acquired a basic understanding of the main EQS elements, it is time to delve deeper and begin implementing a fully working and effective AI agent with it.

Setting up an environment query

In this section, you'll be learning how to add a query to a behavior tree; in particular, we will be tweaking the dummy gunner AI brain in order to let it shoot at a target achieved by an environment query.

> **Note**
>
> As previously mentioned, the EQS is still an experimental feature, so you should use caution if you want to develop a game using it. At the time of writing this book, the EQS is enabled by default by using the **Environment Query Editor** plugin.

Creating the gym

As a first step, we are going to create a proper gym, so start by doing the following:

1. Create a level of your choice, starting from the Level Instances and Packed Level Actors, which I provided in the project template.

2. Add a **BP_GunnerDummyCharacter** instance; just remember to check **Use Controller Rotation Yaw** so it will be able to rotate and point to the target when prompted to do this.

3. Add one or more **BP_Target** instances so that your AI character will have a line of sight to them.

4. Add some obstacles that will block the line of sight with the AI agent. The final gym should look similar to the one depicted in *Figure 11.1*:

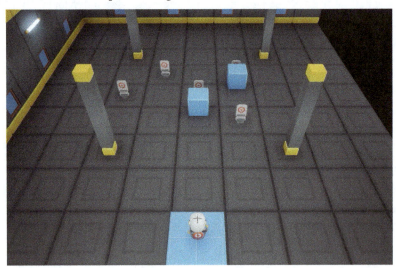

Figure 11.1 – Finished gym

It's now time to set up the AI controller for the AI agent.

Creating the AI controller

The second step is to create a dedicated behavior tree and an AI controller, so do the following:

1. In the `AI` folder, create a new behavior tree and call it `BT_EQSGunnerDummy`.

2. In the `Blueprints` folder, create a new Blueprint class extending from **BaseDummyAIController**, name it `AIEQSGunnerDummyController`, and open it.

3. In the **Details** panel, look for the **Dummy AI Controller** category and set the **Behavior Tree** attribute to **BT_EQSGunnerDummy**.

4. Select the gunner dummy in the level and set its **AI Controller Class** attribute to **AIEQSGunnerDummyController**.

Now that the AI controller and its controller character are properly set, it's time to set up an environment query that we will be using in the behavior tree.

Creating an environment query

We are now going to create an environment query that will look for a viable target in the level. This is going to be pretty similar to the **FindAvailableTarget** task we implemented in *Chapter 9, Extending Behavior Trees*, with a couple of differences; we will be searching for instances of a particular class and the target will need to be in line of sight with the gunner. Additionally, we won't be writing a single line of code. So, let's start by doing the following steps:

1. In the `AI` folder, right-right click and select **Artificial Intelligence | Environment Query** to create one such asset.

2. Name the newly created asset `EQS_FindTarget` and open it.

   You will be presented with a graph named **Query Graph** (pretty similar to the behavior tree one) that will let you implement your own query.

3. Click and drag from the **ROOT** node and add an **Actors of Class** generator node.

4. Select the newly created node and, in the **Details** panel, do the following:

   - Set the **Searched Actor Class** attribute to **BP_Target**

   - Set the **Search Radius** attribute to `3000.0`

5. Right-click on the **ActorsOfClass** node and select **Add Test | Trace**; a test will be added inside the node.

6. Select the test and, in the **Details** panel, do the following:

   - Set the **Test Purpose** attribute to **Filter Only**

   - Set the **Item Height Offset** attribute to `50.0`

   - Set the **Context Height Offset** attribute to `50.0`

7.   Uncheck the **Bool Match** attribute.

The complete graph for the environment query is depicted in *Figure 11.2*:

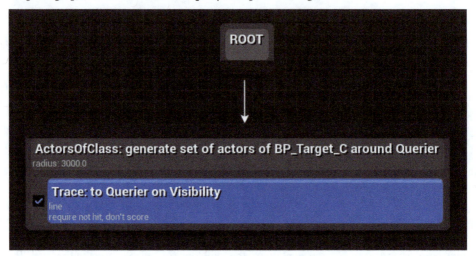

Figure 11.2 – Finished environment query

What we have done here obviously needs some explanation; we have added a generator that will look for all actors of a certain class – **BP_Target** – that are near the querier (the AI agent executing this environment query). Only the items that are in line of sight with the queries will pass the trace test and so they will be the only ones that will be considered for selecting a viable target. **Test Purpose** has been set to **Filter Only** because we only need to have a list of Items and are not interested in their importance – or **score** – in the search.

Now that the environment query has been defined, we can start implementing the behavior tree.

Handling environment queries within a behavior tree

We are now ready to implement the behavior tree, which, as you will see, shows some similarities with the **BT_GunnerDummy** asset; the only difference is the way we get possible targets. For this reason, we will be using the same Blackboard as **BT_GunnerDummy**. So, without further ado, open **BT_EQSGunnerDummy** and, in the **Details** panel, set the **Blackboard Asset** attribute to **BB_GunnerDummy**.

Now, focus on the behavior tree graph and do the following steps:

1.   Add a **Selector** node to the **ROOT** node and name it Root Selector.

2.   Add a **Sequence** node to the **Root Selector** node and name it Shoot Sequence.

3. Add a **Blackboard** decorator to the **Shoot Sequence** node and name it `Has Ammo?`. In the **Details** panel, do the following:

   - Set the **Notify Observers** attribute to **On Value Change**

   - Set the **Key Query** attribute to **Is Not Set**

   - Set the **Blackboard Key** attribute to **NeedsReload**

The graph so far should look like the one shown in *Figure 11.3*:

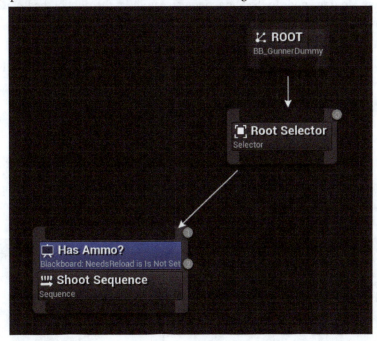

Figure 11.3 – Starting graph

Now, let's connect the environment query to the graph.

4. Add a **Run EQSQuery** task to **Shoot Sequence** and name it `Find Visible Target`.

5. With the task selected, do the following steps:

   - Set the **Query Template** attribute to **EQS_FindTarget**.

   - Set the **Run Mode** attribute to **Single Random Item from Best 25%**.

   - Set the **Blackboard Key** attribute to **TargetActor**.

What we are doing here is executing the environment query and, once the resulting items are returned, we select a random one and assign it to the **TargetActor** key of the Blackboard so it will be available to the behavior tree. Now, let's continue our AI logic.

6. Add **RotateToFaceBBEntry** to **Shoot Sequence** at the right of the **Find Visible Target** node and name it `Rotate Towards Target`.

7. Select the newly created node and set the **Blackboard Key** attribute to **TargetActor**.

   You should be already familiar with this portion of the graph; we are simply rotating the AI agent toward the target, in order to be ready to shoot.

   Let's go on with the graph.

8. Add a **Play Montage** node to the **Shoot Sequence** node, at the right of the **Rotate Towards Target** node and name it `Shoot Montage`.

9. Set the **Anim Montage** property to **AM_1H_Shoot**.

10. Add a **Set Ammo** service to the node and name it `Deplete Ammo`. In the **Details** panel, do the following steps:

    - Set the **Needs Reload** property to **NeedsReload**

    - Check the **Key Value** property

    This will start the shooting animation montage and, subsequently, will spawn a projectile.

11. Add a **Wait** task to **Shoot Sequence** at the right of the **Shoot Montage** node, setting the **Wait Time** attribute to `2.0` and the **Random Deviation** attribute to `0.5`. This portion of the graph should look like the one depicted in *Figure 11.4*:

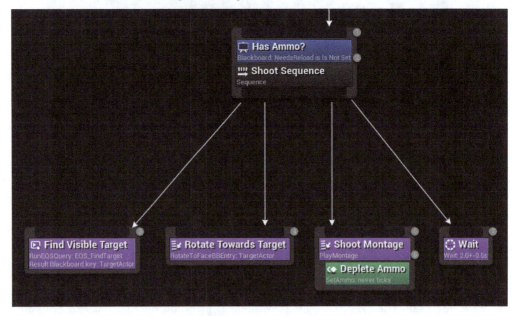

Figure 11.4 – Shoot sequence

We now need to create the reload sequence.

12. Add a **Sequence** node to **Root Selector** at the right of **Shoot Sequence** and name it `Reload Sequence`.

13. Add a **Play Montage** node to the **Reload Sequence** node and name it `Reload Montage`.

14. Set the **Anim Montage** property to **AM_1H_Reload**.

15. Add a **Set Ammo** service to the node and name it `Refill Ammo`. In the **Details** panel, do the following steps:

    - Set the **Needs Reload** property to **NeedsReload**

    - Leave the **Key Value** property unchecked

16. Add a **Wait** task to **Reload Sequence** at the right of the **Reload Montage** node, setting the **Wait Time** attribute to `2.0` and the **Random Deviation** attribute to `0.5`. This portion of the graph is depicted in *Figure 11.5*:

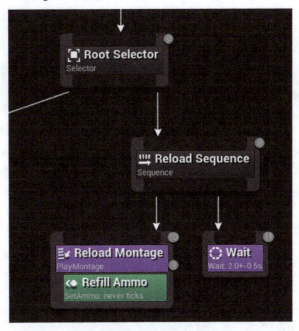

Figure 11.5 – Reload sequence

The behavior tree is now finished and ready to be used.

In this section, we showed how to integrate an environment query inside a behavior tree; in the next section, we will be testing its functionality, and I will show you how to use the debugging tools with it.

Displaying EQS information

To test your gym, just start the simulation; you should notice the following behavior:

- The AI agent will shoot at a target and then reload the weapon

- The AI agent will avoid shooting at targets that are not in the line of sight

This behavior should go on indefinitely; that's perfectly fine as the gunner is looking for targets by searching for the BP_Target class. To improve the gym, you may wish to implement some extra logic, such as a destroy system for targets that have been hit or a points counter for each target that has been hit.

To show on the screen what's happening, you will need to enable the debugging tools, as explained in *Chapter 6, Optimizing the Navigation System*. Once the debugging tools are enabled, you will just need to hit the *3* key on your numpad to show the EQS information and then select the AI agent; in our case, you should see a set of information similar to the one depicted in *Figure 11.6*:

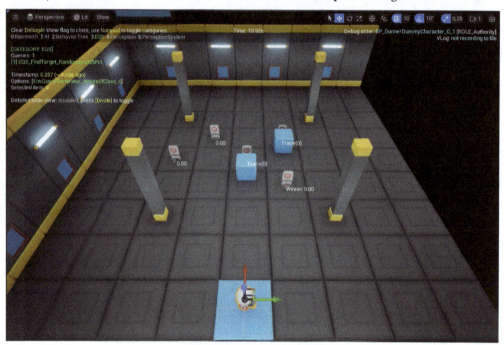

Figure 11.6 – EQS debugging

Obviously, each environment query will have its own information. In this example, you will notice that each target will show some data and a red wireframe, and the selected target will be marked as a **Winner**, as shown in *Figure 11.7*:

Figure 11.7 – Selected target

Targets that won't be in the line of sight with the querier will be labeled as **Trace(0)** to show that the trace failed, as shown in *Figure 11.8*:

Figure 11.8 – Target not in the line of sight

If you want more detailed information for each item in the query, as shown in *Figure 11.9*, you may want to hit the divide key (/) from the numpad:

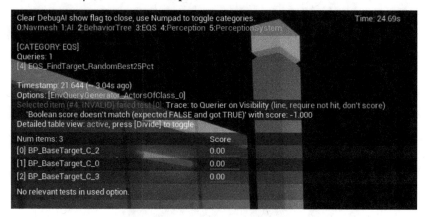

Figure 11.9 – Detailed information

I highly encourage you to create your own queries and test them in the level. Using the debugging tools can be a significant time-saver during the development of your project. Take full advantage of these tools to streamline your workflow and enhance efficiency.

Summary

In this chapter, we saw a new way – although experimental – to let your AI agents probe the level and gather information in order to make meaningful decisions.

Beginning with a bit of theory, we created an environment query, developed a fully operational behavior tree utilizing environment queries effectively, and tested it in a gym setting. Furthermore, we explored the use of debugging tools to analyze the activities within the level.

This concludes *Part 3* of this book. In the upcoming chapters, we will delve into new features within the AI framework, beginning with a glimpse at an alternative method of implementing AI agents: hierarchical state machines.

Get This Book's PDF Version and Exclusive Extras

UNLOCK NOW

Scan the QR code (or go to `packtpub.com/unlock`). Search for this book by name, confirm the edition, and then follow the steps on the page.

Note: Keep your invoice handy. Purchases made directly from Packt don't require an invoice.

Part 4: Exploring Advanced Topics

In the fourth part of this book, you will be presented with some cutting-edge and experimental AI features within the Unreal Engine ecosystem. Additionally, you will gain knowledge on how to integrate these features within your own projects.

This part includes the following chapters:

- *Chapter 12, Using Hierarchical State Machines with State Trees*
- *Chapter 13, Implementing Data-Oriented Calculations with Mass*
- *Chapter 14, Implementing Interactable Elements with Smart Objects*

12

Using Hierarchical State Machines with State Trees

Unreal Engine provides a powerful framework for creating complex AI behaviors through **hierarchical state machines** called **state trees**. By defining various states and transitions, developers can design sophisticated AI logic that adapts to dynamic environments. State trees offer a structured approach to managing AI behavior, allowing for efficient decision-making and seamless integration with other systems in Unreal Engine. What's more, with state trees, you can build clever AI agents that respond to environmental stimuli and interact with the game world in a natural and realistic manner. The purpose of the chapter is to introduce you to the state tree framework in Unreal Engine and to its basic concepts.

In this chapter, we will have a quick introduction to the state tree system available in Unreal Engine and see how to implement state trees inside a project.

In this chapter, we will be covering the following topics:

- Introducing state trees
- Creating and managing state trees
- Using advanced state tree features

Technical requirements

To follow the topics presented in this chapter, you should have completed the previous ones and understood their content.

You'll be using the starter content available in this book's companion repository located at https://github.com/PacktPublishing/Artificial-Intelligence-in-Unreal-Engine-5. Through this link, locate the section for this chapter and download the following .zip file: Unreal Agility Arena - Starter Content.

Although not mandatory, you can use the code created so far or download the files that correspond to the end of the last chapter by clicking the `Unreal Agility Arena - Chapter 11 - End` link.

Introducing state trees

It appeared inevitable that, sooner or later, someone would uncover Dr. Markus' secret experiments:

Hidden in their secret laboratory, Dr. Markus and Professor Viktoria kept on with their groundbreaking experiments. However, news of their remarkable inventions began to spread like wildfire; paparazzi and curious individuals started flocking to the area, eager to uncover the secrets hidden within the laboratory's walls.

Dr. Markus and Professor Viktoria realized they needed to take drastic measures to protect their precious research. With their expertise, the ingenious duo started enhancing their beloved AI dummy puppets with advanced algorithms and behavioral patterns. They programmed the puppets to detect and respond to unauthorized intrusions, trying to ensure the safety of their laboratory and research.

State trees are the Unreal Engine version of hierarchical state machines that merge selectors from behavior trees with state machines, enabling users to build efficient and well-organized logic.

> **Note**
>
> A hierarchical state machine is a design pattern used in software development to model complex systems with multiple states and transitions. It extends the concept of traditional finite-state machines by introducing the idea of hierarchically nested states. In a hierarchical state machine, states can be organized into a hierarchical structure, where higher-level states encapsulate and control lower-level states; this nesting allows for a more modular and organized representation of the system behavior. Each state can have its own set of substates, which can further have their own substates, forming a hierarchical tree-like structure. The main advantage of this pattern is that it provides a way to reuse behavior across multiple states. Instead of duplicating similar logic in different states, common behavior can be defined at higher-level states and inherited by their substates. This promotes code reusability, reduces redundancy, and simplifies the overall design.

A state tree is structured hierarchically, with the state selection process generally starting at the root. However, state selection can be initiated from any node within the tree.

When selecting a **state**, the system evaluates a set of **enter conditions** for the state itself; if conditions are met, the selection progresses to the child states. If no child states exist, it means a leaf has been reached and the current state is activated.

Activating a state triggers all states from the root to the leaf state, with each of these states comprising **tasks** and **transitions**.

Upon selecting a state, the chosen state and all its parent states become active, executing tasks for all active states starting from the root down to the leaf state. All tasks in a state are executed concurrently, and the first task reaching completion will trigger a transition that may result in the selection of a new state.

Transitions can point to any state in the tree, and they are triggered by a set of **trigger conditions** that must be satisfied for the transition to proceed. *Figure 12.1* shows a typical example of a state tree (example taken from the **City Sample** project freely available in the Epic Games Marketplace):

Figure 12.1 – State tree example

To sum it up, the main elements of a state tree are as follows:

- **Root** state: the first state selected when the state tree starts executing its logic
- **Selector** state: state with child states
- **State enter condition**: lists the conditions that decide if a state can be selected
- **Task**: lists a set of actions that will be executed when a state is activated
- **Transition**: the conditions that will trigger the state selection process

> **Note**
>
> In case you are wondering, state trees and behavior trees are both decision-making architectures used in AI, but they serve different purposes. State trees are structured around discrete states and transitions, focusing on the current state of an entity and how it changes in response to events. This makes them suitable for scenarios where clear, distinct states are necessary. In contrast, behavior trees are designed for more complex and fluid decision-making, allowing for modular and hierarchical task execution. They enable smoother transitions between tasks and can handle more intricate behaviors by combining simple actions into complex sequences.

Now that you have a basic understanding of the main state tree terminology, we will show how to extend your own state trees.

Extending state trees

A state tree can be created to be executed on an AI controller or directly from an actor. There are two different components available to handle a state tree:

- `StateTreeComponent`: This can be attached to any actor and be executed by the actor itself
- `StateTreeAIComponent`: This can be attached to any AI controller and be executed by the AI controller itself

Additionally, as you may have already guessed, the state tree system has been created with extensibility in mind, which means you can create your own tasks, evaluators, and conditions. Although you can create your own C++ structures, state trees have been implemented with Blueprint creation in mind. In particular, the main classes available are as follows:

- `StateTreeTaskBlueprintBase`: Used for implementing your own tasks
- `StateTreeEvaluatorBlueprintBase`: Used for implementing your own evaluators
- `StateTreeConditionBlueprintBase`: Used for implementing your own conditions

When extending state trees, it's advisable to implement your own node logic using Blueprints rather than C++. This approach can enhance flexibility and ease of use.

Understanding the state tree flow

State selection in a state tree starts from the tree root and continues down the tree by evaluating each enter conditions. The evaluation process follows these steps:

- If enter conditions are not satisfied, selection goes to the next sibling state
- If enter conditions are satisfied and the state is a leaf, it is selected as the new state
- If enter conditions are satisfied but the state is not a leaf, the process is executed for the first child state

It should be noted that state selection is run dynamically, triggered by transitions. During the first tick, there is an implicit transition to the root state, which then determines the initial state to be executed. Subsequently, once this state is chosen, the transitions specify the conditions that trigger the selection logic, determining when and where it will be executed.

Once a state is selected, all its tasks are executed and will keep on executing until a transition triggers a new state selection process.

Data binding

In game programming, **data binding** refers to the process of connecting data between different parts of the game – such as the user interface and the game logic – and it involves creating a link that allows data to be synchronized and updated across various elements of the game. This helps in keeping the game elements consistent and up to date with the latest information.

State trees use data binding for transferring data within the tree and to establish conditions or configure tasks for execution. Data binding allows access to data passed into the state tree or between nodes in a specified manner.

State tree nodes have the following elements available to implement data binding:

- **Parameters**: These can be referenced during the tree's execution
- **Context data**: This represents predefined data available to the state tree
- **Evaluators**: These are separate classes that can be executed at runtime and that expose data that could not be made available with parameters and context data
- **Global tasks**: These are executed before the root state and can be used when you need permanent data during state selection

It is also worth mentioning that nodes in a state tree share data among themselves and can bind data in three ways that have been previously mentioned:

- State enter conditions
- Transition conditions
- Tasks

In this section, we introduced state trees and their key features. In the following section, we will delve into practical exercises by crafting our own state trees in order to use them effectively within a gym setting.

Creating and managing state trees

From this section onward, we will be creating a new gym based on a couple of AI agents using state trees instead of behavior trees. This will help us to understand the basic principles behind this new development pattern.

To help us understand these principles, we'll be doing the following:

- Create/place an actor in the level who will periodically emit a noise by using a dedicated state tree
- Create/place a dummy character who will be managed by another state tree and will do the following:
  - Stay idle in its starting location
  - Reach the noise location whenever a noise is perceived
  - Get back to its starting location after a brief time investigating the location

Although pretty simple, this logic can be used as a starting point for a guard AI agent that will investigate a level looking for intruders and responding to any suspicious noise around.

As the state trees feature is not enabled by default, the first thing to do will be to get to the **Plugins** window and enable it.

Enabling state trees plugins

To start working with state trees, you will need to enable a couple of dedicated plugins. In order to do so, follow these steps:

1. Open the **Plugin** window by selecting **Edit | Plugins** from the main menu.
2. Search for the **GameplayStateTree** and **StateTree** plugins and enable them.

3.   Restart the Unreal Engine Editor to activate them.

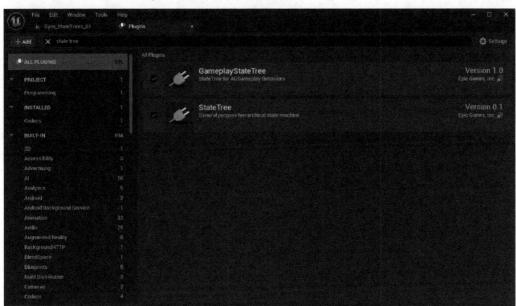

Figure 12.2 – Plugin window

If you plan to use state trees with C++ (and we do), you will need to add a module to the Unreal Engine build file.

In our case, we will need to include the GameplayStateTreeModule dependency in the build, so that the module will be available to your C++ implementation. In order to do so, open your IDE and locate the UnrealAgilityArena.build.cs file in your project; it should be located in the UnrealAgilityArena/Source folder.

> **Note**
> The Unreal Engine .build.cs file is responsible for defining how the project is built, including options for defining module dependencies.

Look for this line of code:

```
PublicDependencyModuleNames.AddRange(new string[] { "Core",
"CoreUObject", "Engine", "InputCore" });
```

Update it to the following:

```
PublicDependencyModuleNames.AddRange(new string[] { "Core",
"CoreUObject", "Engine", "InputCore", "GameplayStateTreeModule" });
```

So, once you declare this module, you will have all that's needed to work with state trees in C++.

> **Note**
>
> At the time of writing this book, there seem to be some issues in declaring a `StateTreeAIComponent` class in C++, while the `StateTreeComponent` class is working just fine. Basically, the `StateTreeAIComponent` class seems to be unavailable in the module, and using this class won't compile your project. To overcome this issue, we will be adding the `StateTreeAIComponent` class from Blueprint when needed.

Once the plugins have been activated, we can start implementing our first AI agent using a state tree: the noise emitter.

Implementing a noise emitter actor

We will be now creating an actor whose sole aim will be to periodically emit a noise by means of the **Perception System**. This task is straightforward and can be implemented in any manner you find suitable. However, for this demonstration, we will use state trees to grasp the fundamental principles of this system.

Creating the noise emitter class

We will be starting by creating the base class for the noise emitter; we will need to declare all the visual elements and, most importantly, the needed Perception System component and the state tree component. Moreover, we will include a function to generate the noise, without concerning ourselves with the logic that will manage it; this responsibility will be delegated to the state tree.

Let's start by creating a new C++ class extending `Actor` and call it `BaseNoiseEmitter`. Once the class has been created, open the `BaseNoiseEmitter.h` file and add the following forward declarations after the `#include` declarations:

```
class UAIPerceptionStimuliSourceComponent;
class UStateTreeComponent;
```

After that, make the class an acceptable base class for Blueprints by changing the `UCLASS()` macro into the following:

```
UCLASS(Blueprintable)
```

Look for the `Tick()` declaration and remove it as we won't be using it.

Next, add the needed components just after the `GENERATED_BODY()` macro:

```
UPROPERTY(VisibleAnywhere, BlueprintReadOnly, Category="Dummy Target",
meta=(AllowPrivateAccess="true"))
UStaticMeshComponent* BaseMeshComponent;

UPROPERTY(VisibleAnywhere, BlueprintReadOnly, Category="Dummy Target",
meta=(AllowPrivateAccess="true"))
UStaticMeshComponent* DummyMeshComponent;

UPROPERTY(VisibleAnywhere, BlueprintReadOnly, Category="Dummy Target",
meta=(AllowPrivateAccess="true"))
UAIPerceptionStimuliSourceComponent* PerceptionStimuliSourceComponent;

UPROPERTY(VisibleAnywhere, BlueprintReadOnly, Category="Dummy Target",
meta=(AllowPrivateAccess="true"))
UStateTreeComponent* StateTreeComponent;
```

As you can see from the preceding code, we will be using a few static meshes, as well as the needed state tree and Perception System components.

Now, just after the constructor declaration, add the following declarations:

```
UPROPERTY(EditAnywhere, BlueprintReadWrite, Category = "Noise
Generation")
float MaxNoiseRange = 3000.f;

UPROPERTY(EditAnywhere, BlueprintReadWrite, Category = "Noise
Generation")
float NoiseRangeRandomDeviation = 100.f;

UPROPERTY(EditAnywhere, BlueprintReadWrite, Category = "Noise
Generation")
FName NoiseTag = "EmitterNoise";

UFUNCTION(BlueprintCallable)
void EmitNoise();
```

As you can see, we are exposing some properties to customize our emitter instances in the level, and we are declaring an `EmitNoise()` function that we will be using to activate the noise emission when needed. Finally, the `NoiseTag` property will be used to tag the noise and be recognized by listening AI agents.

It's now time to open the `BaseNoiseEmitter.cpp` file and implement the methods. As a first step, remove the `Tick()` function and, in the constructor, modify this line of code:

```
PrimaryActorTick.bCanEverTick = true;
```

Modify it into this:

```
PrimaryActorTick.bCanEverTick = false;
```

After that, add the needed #include declarations, so add this block of code at the top of the file:

```
#include "Components/StateTreeComponent.h"
#include "Perception/AIPerceptionStimuliSourceComponent.h"
#include "Perception/AISense_Hearing.h"
```

Now, let's initialize the static mesh components inside the constructor by adding the following code:

```
BaseMeshComponent =
CreateDefaultSubobject<UStaticMeshComponent>(TEXT("BaseMesh"));
RootComponent = BaseMeshComponent;
static ConstructorHelpers::FObjectFinder<UStaticMesh>
  BaseStaticMeshAsset(
    TEXT("/Game/KayKit/SpaceBase/landingpad_large.landingpad_large"));
if (BaseStaticMeshAsset.Succeeded())
{
    BaseMeshComponent->SetStaticMesh(BaseStaticMeshAsset.Object);
}

DummyMeshComponent = CreateDefaultSubobject<UStaticMeshComponent>(TEXT
  ("DummyMesh"));
DummyMeshComponent->SetupAttachment(RootComponent);
DummyMeshComponent->SetRelativeRotation(FRotator
  (0.f, -90.f, 0.f));
DummyMeshComponent->SetRelativeLocation(FVector
  (0.f, 0.f,  80.f));

static ConstructorHelpers::FObjectFinder<UStaticMesh>
  DummyStaticMeshAsset(TEXT("/Game/KayKit/PrototypeBits/Models/Dummy_
    Base.Dummy_Base"));
if (DummyStaticMeshAsset.Succeeded())
{
    DummyMeshComponent->SetStaticMesh(DummyStaticMeshAsset.Object);
}
```

You should be already familiar with all of this, so we can go on and declare the stimuli source for the Perception System and the state tree by adding this piece of code:

```
PerceptionStimuliSourceComponent = CreateDefaultSubobject
  <UAIPerceptionStimuliSourceComponent>(TEXT("PerceptionStimuli
    Source"));
PerceptionStimuliSourceComponent->RegisterForSense
  (UAISense_Hearing::StaticClass());
```

```
StateTreeComponent = CreateDefaultSubobject<UStateTreeComponent>(TEXT
  ("StateTree"));
```

As you can see, we are just creating the needed components; additionally, we are registering the hearing sense as a stimuli source for the Perception System.

Now, in the `BeginPlay()` function, add the following lines of code:

```
PerceptionStimuliSourceComponent->RegisterWithPerceptionSystem();
StateTreeComponent->StartLogic();
```

Here, we are registering the Perception System, and we are starting the logic for the state tree. This means that as soon as the game starts, the state tree will begin executing.

The last thing to do is implement the `EmitNoise()` function, so add this piece of code:

```
void ABaseNoiseEmitter::EmitNoise()
{
    const auto NoiseRange = MaxNoiseRange +
      FMath::RandRange(-1.f * NoiseRangeRandomDeviation,
      NoiseRangeRandomDeviation);
    UAISense_Hearing::ReportNoiseEvent(GetWorld(),
      GetActorLocation(), 1.f, this, NoiseRange, NoiseTag);
}
```

In *Chapter 10, Improving Agents with the Perception System*, you already learned how to handle the sight sense. With the sense of hearing, things are slightly different; while being visible is a continuous occurrence, being heard only occurs when you make noise. This is why we randomize a noise range – based on the previously declared properties – and we use the `ReportNoiseEvent()` function to emit the actual noise.

This class is ready, and we can now focus on the actual state tree creation, starting from a custom task: something that will tell the actor to emit the noise.

Creating the emit noise task

The state tree task we will be creating needs to just tell the `BaseNoiseEmitter` instance to execute the `EmitNoise()` function. This task will be created as a Blueprint class so, inside **Content Browser** of the Unreal Engine Editor, navigate to the `AI` folder and do the following:

1. Create a new Blueprint class extending from **StateTreeTaskBlueprintBase** and name it STT_
    `EmitNoise`. Double-click on it to open it.

2. In the **My Blueprints** panel, hover over the **FUNCTIONS** section and click the **Override**
    dropdown menu that shows up.

3. Select the **Enter State** option, as depicted in *Figure 12.3*:

Figure 12.3 – Enter State function creation

An **Event Enter State** node will be added to the Event Graph; this event will be executed when a new state in the state tree is entered and if the task is part of the active states.

For this state, we need a reference to the owning actor; as previously mentioned, state trees use data binding for communicating. Therefore, we will take advantage of this feature to create the reference. To do this, follow these steps:

4. Create a new variable that is an **Object Reference** of the **BaseNoiseEmitter** type and name it `Actor`.

5. Select the variable and, in the **Details** panel, locate the **Category** attribute and, in the **Input** field, type `Context`, as shown in *Figure 12.4*:

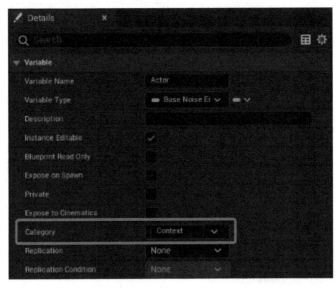

Figure 12.4 – Context category

While variable creation is self-explanatory, setting the category name to a value of **Context** needs some explanation; whenever you add a property to the **Context** category, the property itself will be exposed through data binding to the state tree the task will be executed in. This comes quite handy when you need to get information from the executing state tree and vice versa.

With this new reference available, do the following steps:

6. From the **Variables** section, drag the **Actor** variable into the Event Graph and add a **Get Actor** node.

7. From the outgoing pin of the **Actor** node, connect an **Emit Noise** node.

8. Connect the outgoing execution pin of the **Event Enter State** node with the incoming execution pin of the **Emit Noise** node.

9. From the outgoing execution pin of the **Emit Noise** node, connect a **Finish Task** node.

10. Check the **Succeeded** checkbox of the **Finish Task** node. The final graph should look like the one depicted in *Figure 12.5*:

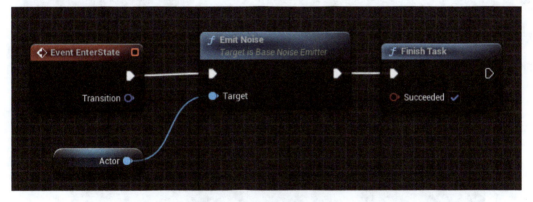

Figure 12.5 – Emit Noise graph

The only thing worth mentioning here is the **Finish Task** node that will return a success value after emitting the noise. Now that this task is complete, we can finally start working on the state tree.

Creating the noise emitter state tree

As previously stated, we will be executing the state tree from an actor; this means we will need something that can be used with the StateTreeComponent class. In order to do so, we will need to create an asset that follows the rules dictated by a StateTreeComponentSchema class that guarantees access to the actor executing the state tree. To create such an asset, do the following steps:

1. In **Content Browser**, open the AI folder, right-click on it, and select **Artificial Intelligence | State Tree**.

2. From the **Pick Schema for State Tree** pop-up window, select **StateTreeComponentSchema**.

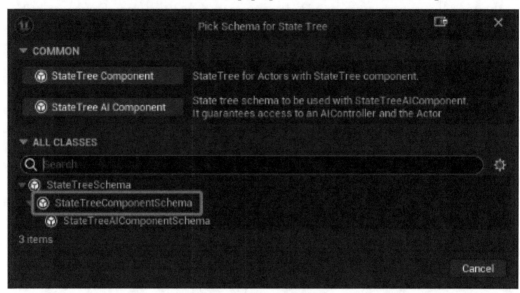

Figure 12.6 – State tree creation

3. Name the newly created asset `ST_NoiseEmitter` and double-click on it to open it.

    Once the asset is opened, locate the **StateTree** tab to the left of the Editor and notice that there is a **Context Actor Class** property, as shown in *Figure 12.7*:

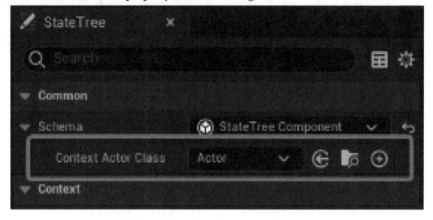

Figure 12.7 – State Tree context actor

This is the reference to the owning actor; at the moment, it is set to a generic actor but we need to be more specific, so click on the dropdown menu and select an **Object Reference** of **BaseNoiseEmitter**. From now on, each node of the tree will be granted access to this reference.

Now, let's start implementing the state tree logic:

1. Click the **Add State** button three times to create three states.

2. Select each of them and, in the **Details** panel, name them Random Delay, Debug Message, and Emit Noise, respectively. Your state tree should be similar to the one depicted in *Figure 12.8*:

Figure 12.8 – Initial states

As you can see, we have created the basis structure for the state tree with three main states that will wait for a random time, display a debug message, and finally emit the noise.

We now need to implement each of the states with their own tasks and transitions; let's start from the first one. Select the **Random Delay** task and do the following:

3. In the **Details** panel, locate the **Tasks** section and click the + button to add a new task.

4. From the newly created task, click the dropdown menu and select **Delay Task**.

5. Expand the task by clicking the tiny arrow icon next to the task name. Set the **Duration** attribute to 10.0 and the **Random Deviation** attribute to 3.0.

6. Locate the **Transitions** section and click the + button to create a new transition.

7. You should see an item labeled **On State Transition Completed Go to State Root**. Click on the tiny arrow icon to expand it, and from the dropdown menu of the **Transition To** attribute, select **Next State**. This state should look like the one depicted in *Figure 12.9*:

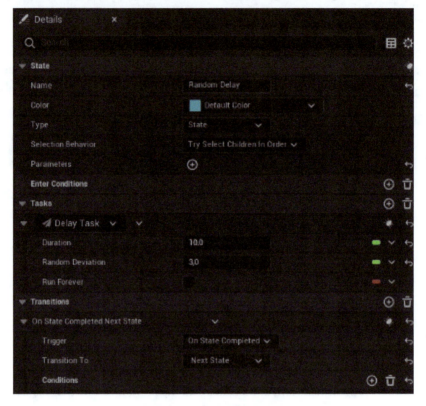

Figure 12.9 – Random delay state

We can now start working on the second state – that is, **Debug Message** – that is going to display an in-game message. This is obviously not mandatory for making our state tree work, but it serves the purpose of learning how things work. Select this state and do the following:

8. In the **Details** panel, locate the **Tasks** section and click the + button to add a new task.

9. From the newly created task, click the dropdown menu and select **Debug Text Task**.

10. Expand the task by clicking the tiny arrow icon next to the task name. Set the **Text** property value to **Emitting Noise!** and the **Text Color** property to a color of your choice – in my case, I opted for a bright yellow.

11. Create another task; click the dropdown menu and select **Delay Task**.

12. Expand the task by clicking the tiny arrow icon next to the task name and set the **Duration** attribute to 0 . 5.

13. Locate the **Transitions** section and click the + button to create a new transition.

14. You should see an item labeled **On State Transition Completed Go to State Root**; click on the tiny arrow icon to expand it and, from the dropdown menu of the **Transition To** attribute, select **Next State**. This state should look like the one depicted in *Figure 12.10*:

Figure 12.10 – Debug Message state

Now select the last state – that is, **Emit Noise** – and do the following:

15. Add a new task, of the **STT Emit Noise** type.

16. Expand the task by clicking the tiny arrow icon; you should notice a property named **Actor** and labeled **CONTEXT**. On its far right, you should see a dropdown menu arrow. Click on it to open it and select **Actor**.

Figure 12.11 – Binding

This last action created the binding between the **Actor** property of the state tree and the **Actor** property we added when we created the **STT_EmitNoise** task. This last property has been exposed because in the task Blueprint; we set its category to **Context**.

This last state will look like the one shown in *Figure 12.12*:

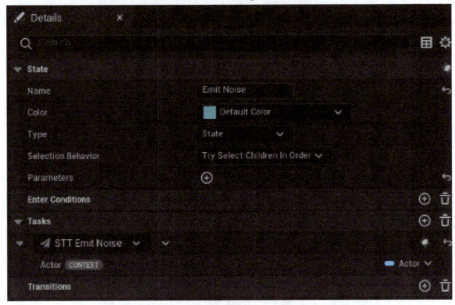

Figure 12.12 – Emit noise state

Please note that we didn't add a transition for this task. The default behavior is to point to the **Root** node, and we want to create an infinite loop, so we simply leave the default behavior.

> **Note**
>
> If you are unfamiliar with data binding, things might appear a bit weird initially, but do not worry. As you become accustomed to it, things will become quite simple and easy to understand.

The state tree is finished and should look like *Figure 12.13*:

Figure 12.13 – Finished state tree

As you can see, it is quite easy to get what's happening in each state and how the state tree flow will progress. It's time to bring everything together and get the noise emitter up and running!

Creating the noise emitter Blueprint

Creating the Blueprint out of the `BaseNoiseEmitter` class is quite straightforward, so let's do the following steps:

1. Create a new Blueprint class extending from **BaseNoiseEmitter** and name it `BP_NoiseEmitter`.

2. Open it and locate the **State Tree** attribute in the **AI** section. From the dropdown menu, set its value to **ST_NoiseEmitter**.

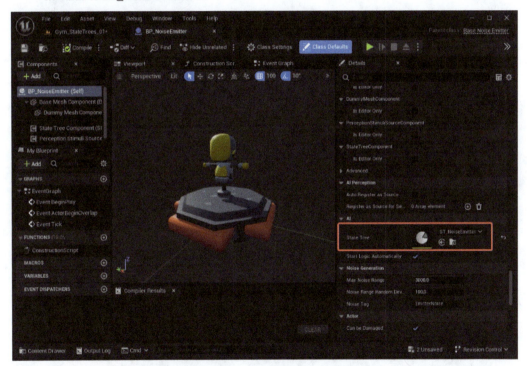

Figure 12.14 – Finished BP_NoiseEmitter

This is all you need to do to implement your noise emitter. It is important to note that what we have created is not technically an AI agent. This is the beauty of state trees; once you grasp the concept, you will be able to apply your logic to many different types of use cases.

Now that the noise emitter is ready, it is time to test it out in a level.

Testing the noise emitter

To test the noise emitter, you will just need to create a new gym and add some instances of **BP_NoiseEmitter**. To do that, start by doing the following:

1. Create a new level, starting from the Level Instances and Packed Level Actors I provided in the project template.

2. Add one or more instances of **BP_NoiseEmitter** to the level.

3. Play the level.

> **Note**
>
> Please note that the **Debug Text Task** message won't be displayed if we hit the **Simulate** button. To show in-game messages, you will need to use the regular **Play** button.

As the level is played, you will see each **BP_NoiseEmitter** instance showing debug messages at random times, as shown in *Figure 12.15*:

Figure 12.15 – Testing the gym

In this section, we saw how to implement a state tree inside an actor. Starting from creating a custom task that calls a method in the owning actor, we then created our first state tree that looped indefinitely, emitting noise signals for the Perception System.

In the following section, we will create a guard puppet that listens for noise events and responds accordingly. We will do this using state trees.

Using advanced state tree features

In this section, we will once again extend the `BaseDummyCharacter` class to create an AI agent that listens for noise signals and moves to the location where the noise was generated. Once the location has been checked, the AI agent will return to its original position. We will start by creating an AI controller that will have hearing capabilities through the Perception System and will handle its AI logic through state trees. We are essentially developing a guard to protect a level from intruders. As usual, let's start by creating our own base C++ class.

Creating the C++ AI controller

The AI controller class will need to implement a hearing sense and execute state tree logic. As mentioned earlier, at the time of writing this book, there seems to be a bug in Unreal Engine that prevents us from declaring a `StateTreeAIComponent` class in C++ so, for the time being, we will be implementing just the hearing sense and adding the state tree component from Blueprints. Let's create a new C++ class called `BaseGuardAIController`. Then, open the `BaseGuardAIController.h` file and add the following forward declaration after the `#include` declarations:

```
struct FAIStimulus;
```

Then, add a `public` section with the following function declarations:

```
public:
    ABaseGuardAIController();

    UFUNCTION(BlueprintCallable,
      BlueprintImplementableEvent)
    void OnTargetPerceptionUpdate(AActor* Actor,
      FAIStimulus Stimulus);
```

We are already familiar with these function declarations but note that the `OnTargetPerceptionUpdate()` function has a `BlueprintImplementableEvent` specifier added; this will let us implement this function from an extending Blueprint instead of directly doing it from this class. This means we are leaving the responsibility of implementing this function to the Blueprint. Now, let's open the `BaseGuardAIController.cpp` file to implement the functions. The needed `#include` declarations you should be adding are the following:

```
#include "Perception/AIPerceptionComponent.h"
#include "Perception/AISenseConfig_Hearing.h"
```

Then, add the constructor implementation:

```
ABaseGuardAIController::ABaseGuardAIController()
{
const auto SenseConfig_Hearing =
  CreateDefaultSubobject<UAISenseConfig_Hearing>
    ("SenseConfig_Hearing");
    SenseConfig_Hearing->
      DetectionByAffiliation.bDetectEnemies = true;
    SenseConfig_Hearing->
      DetectionByAffiliation.bDetectNeutrals = true;
    SenseConfig_Hearing->
      DetectionByAffiliation.bDetectFriendlies = true;
    SenseConfig_Hearing->HearingRange = 2500.f;
    SenseConfig_Hearing->SetStartsEnabled(true);


    PerceptionComponent =
      CreateDefaultSubobject<UAIPerceptionComponent>(TEXT
      ("Perception"));
    PerceptionComponent->
      ConfigureSense(*SenseConfig_Hearing);
    PerceptionComponent->
      SetDominantSense(SenseConfig_Hearing->
        GetSenseImplementation());
    PerceptionComponent->
      OnTargetPerceptionUpdated.AddDynamic(this,
        &ABaseGuardAIController::OnTargetPerceptionUpdate);
}
```

We already know how to configure an `AIPerceptionComponent` from *Chapter 10, Improving Agents with the Perception System*, so I will not bother you with extra details. Just pay attention to the final line of code, where we are registering the delegate that will manage the hearing stimuli.

You can now compile your project to make this class available to the Blueprint system as, in the next few steps, we will create the AI controller Blueprint.

Implementing the AI controller Blueprint

Now that the base AI controller is ready, we can start implementing a Blueprint version that will also manage the state tree. To get started, in the `Blueprints` folder, create a new Blueprint class extending from **BaseGuardAIController**, call it `AIGuardController`, and open it. Then, do the following steps:

1. Create a new variable of the **Name** type and call it `NoiseTag`. Make it **Instance Editable**. After compiling this Blueprint, set the default value for this variable to **EmitterNoise**.

2. Create another variable of the **Vector** type and name it `NoiseLocation`; make it **Instance Editable**.

3. Create a third variable of the **Vector** type and name it `StartLocation`; make it **Instance Editable**.

   Now, let's handle the state tree by following the steps that ensue.

4. Add a new component of the **StateTreeAI** type.

5. Drag the component into the Event Graph to add a reference to the component itself.

6. From the outgoing pin of the **State Tree AI** node, connect a **Start Logic** node.

7. Connect the outgoing execution pin of the **Event Begin Play** node to the incoming execution pin of the **Start Logic** node. The Event Graph should look like *Figure 12.16*:

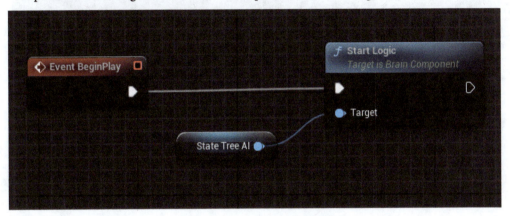

Figure 12.16 – Event Begin Play

This portion of the graph will start the execution of the state tree. Now, let's store the AI character's starting location. To do this, follow the steps that ensue.

8. From the **Variables** section, drag the **StartLocation** variable into the Event Graph and make it a setter node.

9. Connect the outgoing execution pin of the **Start Logic** node to the incoming execution pin of the **Set Start Location** node.

10. Add a **Get Actor Location** node and add its **Return Value** pin to the **Start Location** pin of the **Set Start Location** node.

11. Add a **Get Controller Pawn** node and connect its **Return Value** pin to the **Target** pin of the **Get Actor Location** node. This portion of the graph is depicted in *Figure 12.17*:

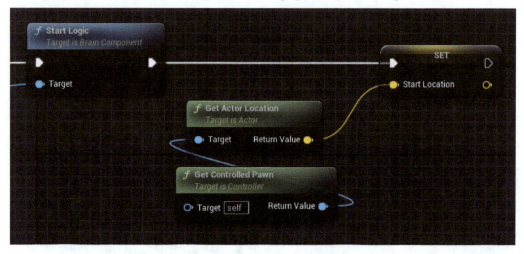

Figure 12.17 – Store starting location

The **Event BeginPlay** code logic has been completed, so we can start implementing the previously declared `OnTargetPerceptionUpdated()` function. To do this, follow these steps:

12. Right-click in the Event Graph and add an **Event On Target Perception Update** node.

13. Click and drag from the **Stimulus** outgoing pin and add a **Break AIStimulus** node.

14. From the **Variables** panel, drag a getter node for the **NoiseTag** variable.

15. From the outgoing pin of the **Noise Tag** node, add an **Equal (==)** node.

16. Connect the **Tag** outgoing pin of the **Break AIStimulus** node to the second incoming pin of the **Equal (==)** node.

17. Connect the outgoing execution pin of the **Event On Target Perception Update** node to a **Branch** node.

18. Connect the outgoing pin of the **Equal (==)** node to the **Condition** pin of the **Break** node. This portion of the graph is depicted in *Figure 12.18*:

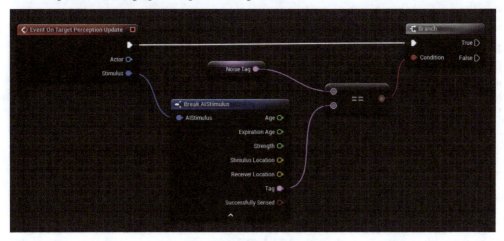

Figure 12.18 – Check stimulus tag

We now need to store the location of the noise, so let's do the following steps:

19. From the **Variables** section, drag the **NoiseLocation** reference to create a **Set** node.

20. Connect the node's incoming execution pin to the **True** execution pin of the **Branch** node.

21. Connect the **Noise Location** pin of the **Set Noise Location** node to the **Stimulus Location** pin of the **Break AI Stimulus** node. This portion of the graph is shown in *Figure 12.19*:

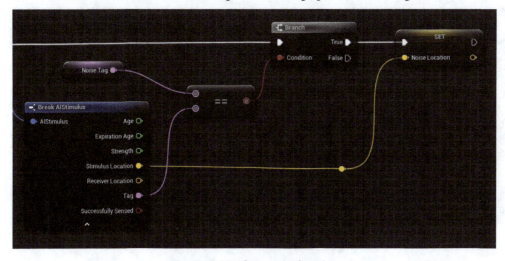

Figure 12.19 – Store noise location

The last thing we need to do is to notify the state tree that a noise has been heard and that the AI agent needs to respond consequently. To do so, we will be using **Gameplay Tags**.

> **Note**
>
> In Unreal Engine, the Gameplay Tag system is used to mark and categorize gameplay elements. Gameplay Tags are lightweight identifiers that can be easily attached to game entities (such as actors or components) to help organize and classify them in a flexible and efficient way. Learning how to work with Gameplay Tags is out of the scope of this book; we will be just learning the bare minimum to properly communicate with a state tree.

Let's get on with our Event Graph implementation by doing the following steps:

22. Add a **Make StateTreeEvent** node to the graph.

23. Add a **Send State Tree Event** node to the graph.

24. Drag a reference of the **StateTreeAI** component into the graph.

25. Connect the outgoing execution pin of the **Set Noise Location** node to the incoming execution pin of the **Send State Tree Event** node.

26. Connect the **State Tree AI** node to the **Target** pin of the **Send State Tree Event** node.

27. Connect the outgoing pin of the **Make State Tree Event** node to the incoming **Event** pin of the **Send State Tree Event** node.

28. In the **Origin** input field of the **Make StateTreeEvent** node, type AI Controller. This portion of the graph should look like *Figure 12.20*:

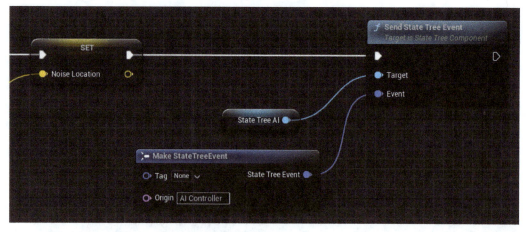

Figure 12.20 – Send State Tree Event

This portion of the code is responsible for communicating with the state tree by sending an event; the event will need to be tagged in order to be recognized by the state tree itself. To do so, we need to create a Gameplay Tag. You can do this by following these steps:

29. In the **Make StateTreeEvent** node, click on the dropdown menu next to the **Tag** incoming pin. At the moment, it should be labeled as **None**.

30. You will get a list of available tags; click on the **Manage Gameplay Tags…** option to open the **Gameplay Tag Manager** window.

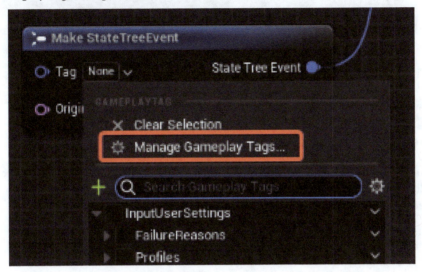

Figure 12.21 – Manage Gameplay Tags… option

31. Once the **Gameplay Tag Manager** window is open, click on the + button. In the **Name** input field, enter UnrealAgilityArena.StateTree.HeardNoise and, in the **Source** field, select **DefaultGameplayTags.ini**.

32. Click the **Add New Tag** button to confirm the creation of the new Gameplay Tag.

Figure 12.22 – Create Gameplay Tag

33. In the **Make StateTreeEvent** node, click on the **Tag** dropdown menu and select the **HeardNoise** checkbox to select that gameplay tag.

Figure 12.23 – Select Gameplay Tag

We're almost done with the **AIGuardController** Blueprint; the only thing left is to include the state tree reference, but we have to create it first!

Implementing the state tree

Now, we are going to implement the state tree. This time, we will be executing it from the **AIGuardController** Blueprint. That's why we will need a subclass of the regular state tree – that is the state tree AI – that will have a reference to the owning AI controller.

The main states will be the following:

- **Idle**: The AI agent will stay at its starting location

- **Alerted**: The AI agent has been notified of a noise and it will go inspecting the location

- **Return to Starting Location**: The AI agent will get back to its starting location

So, let's start by opening the AI folder and doing the following steps:

1. Right-click and select **Artificial Intelligence | State Tree**.

2. From the **Pick Schema for State Tree** pop-up window, select **StateTreeAIComponentSchema** and name the newly created asset STAI_Guard. Double-click on it to open it.

3. In the **StateTree** left panel, set the **AIController Class** attribute to **AIGuardController** and the **Context Actor Class** attribute to **BP_Guard**.

As we have already mentioned, the aforementioned steps will bind the state tree to the owning AI controller and actor; this way you will be granted access to their properties.

We are now going to implement the base states. To do so, carry out the following steps:

1. Create three states and call them, Idle, Warned, Resume, respectively.

2. Select both the **Warned** and **Resume** states and, in the **Details** panel, set the **Type** attribute to **Group**.

We marked the **Warned** and **Resume** states as groups because they won't contain tasks but will delegate them to child states. They basically function as state containers.

As an extra option, the **State Tree** panel has a **Theme** section that will let you define state colors that can then be applied to each state – and its own children – in the **Details** panel. In my case, I have opted for the colors depicted in *Figure 12.24*:

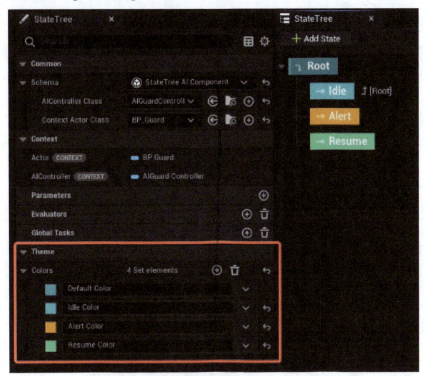

Figure 12.24 – Base states

Let's now implement each state separately.

Implementing the Idle state

The **Idle** state is going to be quite simple; we will make the AI agent wait in some sort of infinite loop until we get a noise notification. To implement this state, select it and do the following steps:

1. Add a **Delay Task** with the **Duration** attribute set to 10.0.

2. Add a **Transition** with the following settings:

   - The **Trigger** attribute set to **On State Completed**

   - The **Transition To** attribute set to **Idle**

3.   Add another **Transition** with the following settings:

- The **Trigger** attribute set to **On Event**

- The **Event Tag** attribute set to **UnrealAgilityArena.StateTree.HeardNoise**

- The **Transition To** attribute set to **Alert**

- The **Priority** attribute set to **High**

Figure 12.25 – Idle state

As you can see, we will keep on looping inside this state until we get an event from the AI controller notifying us that a noise has been heard. In this case, we will transition to the **Alert** state.

Implementing the Alert state

Once in the **Alert** state, the AI agent will try to move to the noise location. Once it reaches that point, will wait for some time before changing state; we will need two child states for this.

So, to create the first child state, do the following steps:

1.   Create a new state and call it Move to Noise Location.

2. Add a task of the **Action | Move To** type and do the following steps with it:

    I.   Bind the **AIController** attribute – labeled **CONTEXT** – to the owner AI controller by clicking the dropdown arrow and selecting **AIController**.

    II. Locate the **Destination** attribute, click the dropdown arrow, and select **AIController | Noise Location** to bind this attribute to the **NoiseLocation** property of the AI controller owner.

3. Add a **Transition** with the following settings:

    - The **Trigger** attribute set to **On State Succeeded**
    - The **Transition To** attribute set to **Next State**

4. Add another **Transition** with the following settings:

    - The **Trigger** attribute set to **On State Failed**
    - The **Transition To** attribute set to **Resume**

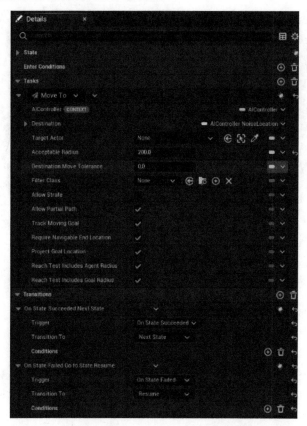

Figure 12.26 – Move to Noise Location state

5. This portion of the state tree will move the AI agent to the location set inside the **NoiseLocation** attribute of the owning AI controller. Once successful, the next state will be executed. If the location cannot be reached, it will get back to its original position.

Let's now create the second child state for the **Alert** state by doing the following steps:

1. Create a new state and call it `Inspect Noise Location`.

2. Add a task of the **Delay Task** type and do the following with it:

    *   Set the **Duration** attribute to `3.0`

    *   Set the **Random Deviation** attribute to `1.0`

3. Add a **Transition** with the following settings:

    *   The **Trigger** attribute set to **On State Completed**

    *   The **Transition To** attribute set to **Resume**

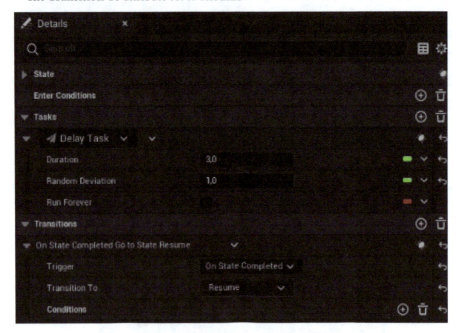

Figure 12.27 – Inspect Noise Location state

This portion of the state tree will just make the AI agent wait for a bit, before it goes back to its original location. If the AI agent doesn't find anything suspicious, it gets back to its guard location.

Implementing the Resume state

The **Resume** state will need to bring the AI agent back to its original position; additionally, at any time, this state should be interrupted if a new noise has been notified. So, to create the first child state, do the following steps:

1. Create a new state and call it `Move to Starting Location`.

2. Add a task of the **Action | Move To** type and do the following steps with it:

   I. Bind the **AI Controller** attribute – labeled **CONTEXT** – to the owner AI controller by clicking the dropdown arrow and selecting **AI Controller**.

   II. Locate the **Destination** attribute, click the dropdown arrow, and select **AIController | Start Location** to bind this attribute to the **StartLocation** property of the AI controller owner.

3. Add a **Transition** with the following settings:

   • The **Trigger** attribute set to **On State Completed**

   • The **Transition To** attribute set to **Next State**

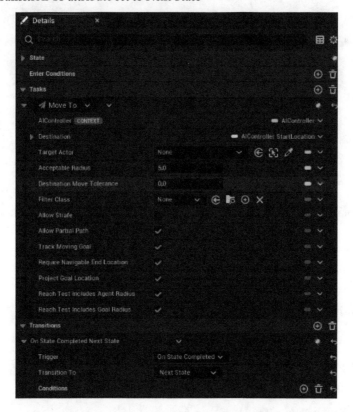

Figure 12.28 – Move to Start Location state

This state is pretty similar to the **Move to Noise Location** one; the only difference is the **Destination** attribute that, in this case, is the original location of the AI agent.

Let's now create the second child state for the **Resume** state by doing the following steps:

1. Create a new state and call it `Wait`.

2. Add a task of the **Delay Task** type and do the following with it:

   • Set the **Duration** attribute to `2.0`

   • Set the **Random Deviation** attribute to `1.0`

3. Add a **Transition** with the following settings:

   • The **Trigger** attribute set to **On State Completed**

   • The **Transition To** attribute set to **Idle**

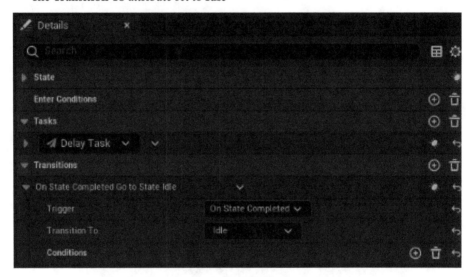

Figure 12.29 – Wait state

As a last step, we will need to interrupt the **Resume** state if a new noise has been heard, so select the **Resume** state and do the following steps:

4. Add a **Transition** with the following settings:

   • The **Trigger** attribute set to **On Event**

   • The **Event Tag** attribute set to **UnrealAgilityArena.StateTree.HeardNoise**

   • The **Transition** To attribute set to **Alert**

   • The **Priority** attribute set to **High**

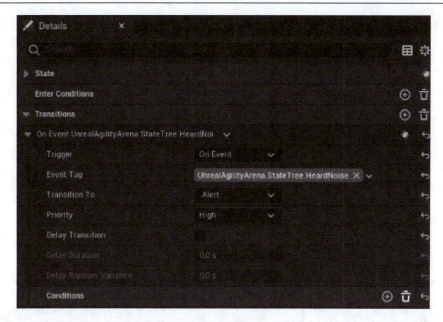

Figure 12.30 – Resume state

The state tree is pretty complete, and it should look like the one shown in *Figure 12.31*:

Figure 12.31 – Finished state tree

Now that the state tree is ready, we will need to add it to the AI controller.

Assigning the state tree to the AI controller

Assigning the newly created state tree to the AI controller is pretty straightforward. Just open the **AIGuardController** Blueprint and, in the **Details** panel, locate the **AI** section. Set the **State Tree** property to **STAI_Guard**.

Once this is done, we can create the AI agent Blueprint.

Creating the guard Blueprint

We will now create our guard Blueprint and assign the AI logic to it. To do this, open **Content Drawer**. In the Blueprints folder, create a new Blueprint extending from **Base Dummy Character** and call it BP_Guard.

Open the newly created Blueprint and, in the **Details** panel, set the **AI Controller Class** attribute to **AIGuardController**.

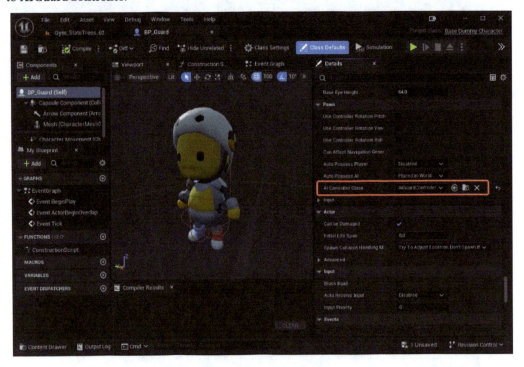

Figure 12.32 – Guard Blueprint

The guard AI agent is now ready; we just need to test it out in a gym.

Testing in a gym

With the Blueprint class ready, it's time to test it out. All you need to do is to add it to the previously created gym and play the level. Whenever a noise is emitted by a noise emitter actor, you should see the **BP_Guard** instance try to reach the noise location and, after a while, get back to its original position. Obviously, all of this will work if the AI agent is in range of the provoked noise location. You can obviously leverage your understanding of AI debugging tools to gain valuable insights into the hearing capabilities and range of the AI agent.

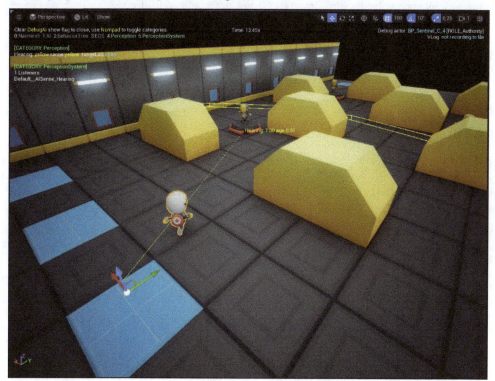

Figure 12.33 – Testing the gym

In this quite lengthy section, you got more advanced information on how to implement your own state trees. Starting from an AI controller with some hearing capabilities, we learned how to control a state tree and bind important data between the state tree and the AI controller. Finally, we added the AI controller and, consequently, the state tree to an AI agent and tested its behaviors in a gym.

The AI agent we have created sets the foundation for a complete AI guard agent. Currently, it simply checks for any suspicious noises and investigates their source. I highly recommend adding your own logic to introduce more actions, such as sounding an alarm when an enemy is detected or trying to attack the source of the noise.

Summary

State trees in Unreal Engine are crucial in the AI framework as they help manage and organize the decision-making processes of AI agents efficiently. They offer a neat alternative to behavior trees depending on your design and development patterns.

In this chapter, we learned the basics of state trees, a hierarchical state machine framework implemented in Unreal Engine. Starting from its main concepts, such as how states are handled, we were introduced to the main elements involved – including tasks, transitions, and conditions. After that, we created our own actors, taking advantage of state trees.

In the next chapter, we will be working on a totally different topic: how to manage huge numbers of objects within your level, treating them as a collective group for processing or simulation purposes.

13

Implementing Data-Oriented Calculations with Mass

The **Mass** framework is a relatively new system in Unreal Engine that allows developers to efficiently manage and manipulate large numbers of objects within a game environment together. It provides tools and functionalities to handle them effectively, optimize performance, and implement behaviors for AI and gameplay mechanics. The Mass framework is becoming an essential tool for creating game levels that require large numbers of NPCs while maintaining optimal performance levels in projects. Taking advantage of the Mass framework is essential for game developers to create immersive and engaging experiences while maintaining optimal performance.

In this chapter, we will be presenting the basics of Mass and showing you how to create your own Mass systems. In particular, we will be covering the following topics:

- Introducing the Mass framework
- Setting up Mass
- Spawning Blueprints

Technical requirements

To follow the topics presented in this chapter, you should have completed the previous ones and understood their content.

Additionally, if you would prefer to begin with code from the companion repository for this book, you can download the `.zip` project files provided in the project repository: `https://github.com/PacktPublishing/Artificial-Intelligence-in-Unreal-Engine-5`

To download the files from the end of the last chapter, click the `Unreal Agility Arena - Chapter 12 - End` link.

Introducing the Mass framework

The Mass framework is a data-oriented system created for managing high-performance computations on extensive collections of entities. As you already know, in Unreal Engine, the traditional approach involves using actors and components to create level objects. This approach offers great flexibility for combining logic within actors, but as the project grows larger, it often results in data inconsistencies that lead to performance issues. For example, consider a large online multiplayer game where different AI agents can perform various actions based on complex logic. Initially, this flexibility allows developers to easily implement features such as character interactions and item trading with minimal constraints. However, as more characters and interactions are added, inconsistencies with data updates across the network can arise.

Mass, on the other hand, employs a data-oriented design framework that offers an alternative data storage method to separate data from processing logic. This allows for easy management of big – or even huge – numbers of entities in the level.

> **Note**
>
> In this chapter, I will often use the term **level of detail (LOD)**. In case you are unfamiliar with this term, it is a crucial concept in game development that refers to the technique of managing the complexity of 3D models based on their distance from the camera. The primary goal of LOD is to optimize rendering performance while maintaining visual fidelity. When a player is close to a static mesh, the engine uses a high-detail version of that mesh to ensure it looks sharp and detailed. However, as the player moves further away, the engine can switch to less detailed versions of the mesh, which require fewer resources to render.

In the upcoming subsections, I will give you a quick introduction to the main definitions and elements that make up the Mass framework.

Mass framework plugins

Mass relies on four main plugins:

- **MassEntity**: The framework core plugin that is required to make Mass work
- **MassGameplay**: This plugin manages situations such as interaction within the world, movement, object visualization, LOD, and so on
- **MassAI**: This plugin manages features such as state trees, world navigation, and avoidance
- **MassCrowd**: This plugin is specialized in handling crowds and has its own specialized visualization and navigation systems

> **Note**
>
> At the time of writing this book, `MassEntity` is in beta and considered production-ready. On the other hand, `MassGameplay`, `MassAI`, and `MassCrowd` are still experimental; this means you should handle them with care as things may change as time goes by.

Understanding Mass elements

The primary data structure within `MassEntity` is the **fragment** – an atomic unit of data used in computations – that can be represented for example by transform, velocity, or LOD index. Fragments can be organized into sets, with a specific set instance linked to an ID, forming an **entity**.

Creating an entity resembles class instantiation in object-oriented programming. However, instead of rigidly defining a class and its functionalities, entities are constructed through fragment composition. These composition assemblies are modifiable at runtime.

> **Note**
>
> Fragments and entities are data-only elements and do not contain any logic.

A set of entities with the same composition is called an **archetype**; each archetype comprises various types of fragments arranged in a specific manner. For instance, an archetype might feature a fragment composition with transform and velocity, indicating that all entities linked to this archetype share the same fragment structure.

Entities within an archetype are grouped into memory **chunks**, optimizing the retrieval of fragments belonging to entities of the corresponding archetype from memory for enhanced performance.

A **ChunkFragment** is a fragment connected to a chunk rather than an entity and is used to store chunk-specific data that is utilized in processing. ChunkFragments are an integral part of an entity's composition.

A **tag** is a simple fragment that does not contain any data; tags are included in an entity's composition.

Processors are stateless classes that provide the processing logic for fragments; by using **EntityQueries**, they specify the types of fragments they require for their operations. Processors can add or remove an entity's fragments or tags, effectively changing the composition.

Fragments and processors delivering specific functionalities are named **traits**; multiple instances of traits can be incorporated into an entity. Each instance of a trait is tasked with integrating and setting up fragments to ensure that the entity displays the behavior associated with that trait. Typical traits include state tree management, avoidance, or debug visualization.

As `MassGameplay` is probably the most important of the Mass plugins, we will focus on it. Let's get into detail on its main features.

MassGameplay

As mentioned, one of the plugins that compose the Mass framework is the `MassGameplay` plugin, which is derived directly from the `MassEntity` plugin. The `MassGameplay` plugin includes a list of powerful subsystems that are listed here:

- **Mass Representation**: This subsystem is responsible for managing different visual aspects of entities

- **Mass Spawner**: This subsystem manages the entities' spawning process

- **Mass LOD**: This subsystem manages the LOD of each mass entity

- **Mass State Tree**: This subsystem will let you integrate state trees inside `MassEntity`

- **Mass Signals**: This subsystem manages messages to entities in a similar way to event dispatching

- **Mass Movement**: This subsystem manages movement for Mass agents

- **Mass SmartObject**: This system is responsible for integrating smart objects – a feature I will be presenting in the next chapter – inside `MassEntity`.

- **Mass Replication**: This subsystem is responsible for replicating Entities over the network in multiplayer games; at the time of writing this book, it is still in the experimental stages

With the availability of all these subsystems, Mass has become an incredibly powerful tool for efficiently managing a vast number of entities.

In this section, we introduced the Mass framework and its main elements. In the next section, we will be creating our first level using Mass, leveraging some of its main features.

Setting up Mass

The story goes on…

As Dr. Markus and Professor Victoria delved deeper into their groundbreaking research on AI, they found themselves faced with a new challenge; the laboratory was now filled with a large number of AI dummy puppets, each designed to mimic human behavior and responses. However, managing and synchronizing these puppets proved to be a daunting task.

In order to overcome this obstacle, Dr. Markus and Professor Victoria started by developing a sophisticated set of algorithms that would act as the core system for managing the puppets. This system would allow them to sync and coordinate the movements and actions of the AI puppets, making them work seamlessly together.

In this section, we will show which plugins are needed to use Mass and how to enable them. Also, we will be working on a new level to get a first peek at the spawning system.

Enabling plugins

In order to make Mass fully functional, you will need to enable the following plugins:

- `MassEntity`
- `MassGameplay`
- `MassAI`
- `MassCrowd`
- `ZoneGraph`
- `StateTree`

`ZoneGraph` and `StateTree` are not part of the Mass framework but need to be included because of some dependencies in the code.

You are already familiar with the **Plugin** window, so open it up and enable the aforementioned plugins. You will get a warning about the experimental features you will be using and then you will be prompted to reload the Unreal Engine Editor in order to register the plugins.

Once this process is finished, you will be ready to get on with Mass.

Creating a MassEntityConfigAsset

In Mass, configuring entities is achieved through an Unreal Engine **data asset**.

In case you are unfamiliar with a data asset, it is a type of asset that stores data in a structured format. It is typically used to store various types of data, such as configuration settings, game parameters, localization text, and more.

In Mass, you will be using a `MassEntityConfigAsset` data type, which will store trait information and fragments. As a starting point, we will be creating a config asset that will just let us debug information on where an object has been spawned in the level. This will let us visualize the current state of the system we are developing and analyze it at runtime.

To get started, in **Content Drawer**, create a new folder named `DataAssets` and open it up. Then, do the following steps:

1. Right-click on **Content Drawer** and select **Miscellaneous | Data Asset**; name the newly created asset `ME_DebugVisualizationConfig`.
2. In the **Config** section, you should see a **Traits** array; hit the + button to create a new item in the array.
3. From the array item dropdown menu, select **Debug Visualization**.
4. Expand the array item by clicking on the arrow.

5. Expand the **Debug Shape** attribute too.

6. In the **Mesh** attribute, search for **Dummy_Base** and select it, as shown in *Figure 13.1*:

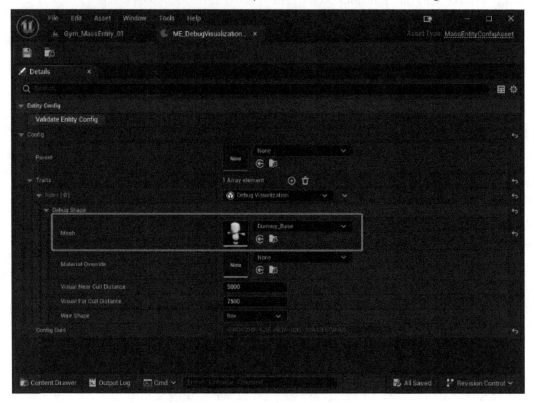

Figure 13.1 – ME_DebugVisualizationConfig

What we have done here is quite simple. We created a `MassEntity` configuration file with a single trait responsible for displaying a debug mesh at the location where an object has been spawned in the level.

> **Note**
>
> If you need to check that everything is fine, you can click the **Validate Entity Config** button that is located inside the **Data Asset** window. You should get a message stating that there are no errors in this asset.

We will now proceed with the object spawn implementation.

Creating a spawn EQS

To take advantage of the Mass Spawner subsystem, we will need an EQS that will return a set of locations that will tell the subsystem where to spawn objects. If you need a refresher about EQS, my advice is to check *Chapter 11, Understanding the Environment Query System*, and then get back to this chapter when you are ready. We will be now creating an environment query that will just generate some points on a grid. In order to do so, open the `AI` folder and follow these steps:

1. Right-click and select **Artificial Intelligence | Environment Query**. Name the newly created asset `EQS_SpawnEntitiesOnGrid`. Double-click on the asset to open it.

2. From the **ROOT** node, click and drag to create a **Points: Grid** node from the **Generators** category.

3. Click on the node and, in the **Details** panel, do the following:

   - Set the **GridHalfSize** attribute to `1500.0`

   - In the **Projection Data** section, set the **Trace Mode** attribute to **Geometry by Channel**

Figure 13.2 – Environment query

The environment query we have just implemented will create a grid sized 3,000 by 3,000 – as we have set the **GridHalfSize** property with a value of `1500.0` units – and will use a **Geometry by Channel** trace method to set the location of each item.

> **Note**
>
> If you are unfamiliar with what traces are and how Unreal Engine uses them, my advice is to take a peek at the official documentation by visiting this page from the official documentation at `https://dev.epicgames.com/documentation/en-us/unreal-engine/traces-in-unreal-engine---overview`.

We are now ready to use the Mass Spawner subsystem inside a level.

Creating the gym

To test the spawning feature, we are going to create a simple gym. To get started, create a level of your choice, starting from the Level Instances and Packed Level Actors I provided in the project template. If you wish, add some obstacles; my level is shown in *Figure 13.3*:

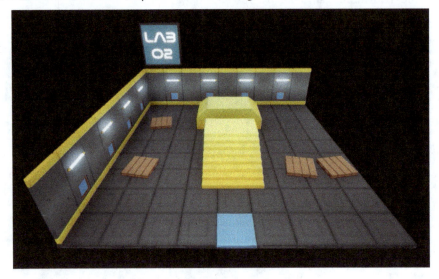

Figure 13.3 – Gym

Now, follow these steps:

1. On the **Editor** toolbar, click on the **Quickly add to the project** button and search for **Mass Spawner**. Select it to add an instance of it in the level.

2. Place the **MassSpawner** object at the center of your level.

3. Now that you have placed the **MassSpawner** object in the level, select it. In the **Details** panel, do the following:

   • Set the **Count** property to 20

   • Click the + button on the **Entity Types** attribute to add an item to the array

- Expand the item marked as **Index[0]** and set the **Entity Config** property to **ME_DebugVisualizationConfig**

Figure 13.4 – Entry types

We just set the **MassSpawner** property so that it spawns 20 entities that will use the data asset we previously created as the configuration asset. We now need to tell the **MassSpawner** property where to spawn them. To do this, follow these steps:

1. In the **Details** panel, add an element to the **Spawn Data Generator** array and expand the **Index[0]** element.

2. Set the **Generator Instance** property to **EQS SpawnPoints Generator**.

3. Expand **Generator Instance** and its **Query** child.

4. Set **Query Template** to **EQS_SpawnEntitiesOnGrid**.

Figure 13.5 – Spawn point generator

In the previous steps, we have set the generator logic to the entity query we had previously created. It's now time to test the gym.

Testing the gym

To test the gym, all you need to do is to simulate – or play – the level; you will see 20 debug models spawned on a grid, as shown in *Figure 13.6*:

Figure 13.6 – Gym

Please note that models will be correctly placed on objects; this is happening because we used the **Geometry by Channel** trace mode inside our environment query.

By enabling the debugging tools, you will have some insightful information about what's happening inside the level along with a plethora of options, as depicted in *Figure 13.7*:

Figure 13.7 – Debugging tools

I highly encourage you to explore the various debugging tool options, as they can greatly enhance your efficiency in identifying and resolving your game issues. As an example, *Figure 13.8* shows the archetype information that was enabled by using the *Shift + A* key combination:

Figure 13.8 – Archetype information

In this section, we got a glimpse of what's possible to create with Mass; in particular, we saw how to create a spawn area through the **MassSpawner** subsystem. In the next section, we will go into more detail by spawning a set of Blueprints instead of some debugging models.

Spawning Blueprints

In this section, we will be taking a step further with the **MassSpawner** subsystem as we will be generating a set of Blueprint instances instead of simply showing debug meshes; this will let us transition from a simple debugging gym to a real case environment. Additionally, we will learn how to handle the LOD for spawned entities.

Let's imagine that we want to create a concert where the audience is automatically generated, and it is managed as a whole by Mass. We will be creating such a scenario using the Mass framework.

Creating the audience Blueprints

As a first step, we will be creating a Blueprint that will serve us as the audience for an imaginary concert; entities will be cheering, sitting, or simply staying still. Additionally, we will be using the Mass LOD management system. Instead of creating just one Blueprint, we will be creating two: one that will be activated by the management system when near the camera and the other that will be activated when far away. This will let us manage and show more complex entities only when near the camera and fall back to less complex entities when far away. For the sake of demonstration, we will be using a Blueprint that will just stay still and won't be animated – the one that will be visualized when far away – and one that will show random animations.

Let's start with the first Blueprint by doing the following steps:

1. Create a new Blueprint class inheriting from **BaseDummyCharacter** and call it `BP_ AudienceLow`.

2. Open the Blueprint, and in the **Animation Mode** property, select **Use Animation Asset**.

3. Uncheck the **Looping** and **Playing** checkboxes.

4. Set the **Initial Position** attribute to `0.4`.

Once visualized in the level, this character will just stand still and show a cheering position.

The second Blueprint is going to be slightly more complex. To implement it, follow these steps:

1. Create a new Blueprint class inheriting from **BaseDummyCharacter** and call it `BP_ AudienceHigh`.

2. Create a new variable of the **Anim Montage Object Reference** type, call it `MontageList`, and make it an array.

3. Create two variables of the **float** type and name them `MinInterval` and `MaxInterval`, respectively.

4. Compile the Blueprint to expose the **Default Value** properties and set the **MinInterval** value to `3.0` and **MaxInterval** to `6.0`.

5. Add three items in **MontageList Default Value** and set them, respectively, to `AM_Cheer`, `AM_Interact`, and `AM_Sit`.

6. In the Event Graph, connect the **Event Begin Play** execution pin to a **Delay** node.

7. Connect the **Duration** pin of the **Delay** node to a **Random Float in Range** node.

8. From the **Variables** section, drag a **MinInterval** getter node and connect it to the **Min** pin of the **Random Float in Range** node.

9. From the **Variables** section, drag a **MaxInterval** getter node and connect it to the **Max** pin of the **Random Float in Range** node. This portion of the graph is depicted in *Figure 13.8*:

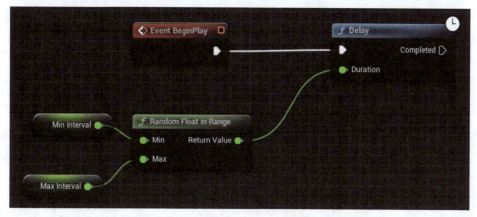

Figure 13.9 – Delay logic

This portion of the graph is nothing fancy and is just adding a random delay to the code logic. Let's get on with the code by doing the following steps:

10. Connect the outgoing execution pin of the **Delay** node to a **Play Montage** node.

11. From the **Components** panel, drag a **Mesh** reference in the Event Graph and connect it to the **In Skeletal Mesh Component** pin of the **Play Montage** node.

12. Drag a **MontageList** getter in the Event Graph and connect its pin to a **Random Array Item** node.

13. Connect the **Out Item** pin of the **Random** node to the **Montage to Play** pin of the **Play Montage** node. This part of the graph is shown in *Figure 13.10*:

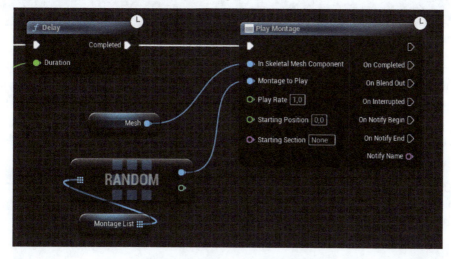

Figure 13.10 – Play Montage logic

Here we are doing something pretty straightforward; we are getting a random animation montage from the array and then we play it.

The last thing to do is to create an infinite loop that will keep on playing an animation montage after a delay. To do this, follow these steps:

14. Connect the **On Completed** outgoing execution pin of the **Play Montage** node to the incoming execution pin of the **Delay** node.

15. Add a couple of reroute nodes, in order to make the graph clearer. The code is shown in *Figure 13.11*:

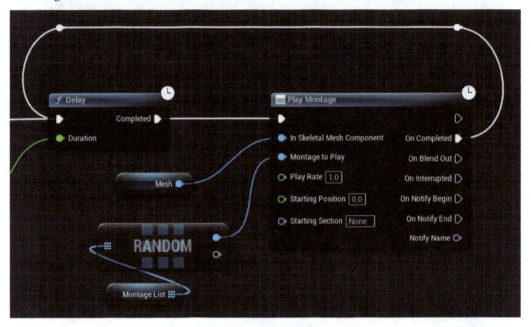

Figure 13.11 – Loop

Our character is now ready and, once instantiated in a level, will keep on playing random animations. We will now create the **MassEntityConfigAsset** for the **MassSpawner** that will be placed in the gym.

Creating a MassEntityConfigAsset

The **MassEntityConfigAsset** we will be creating is going to be more complex than the previous one. In this case, we will need to spawn characters in the level, and we will need to handle what is instantiated depending on the distance from the camera.

To get started, open the `DataAssets` folder, create a new **Data Asset** of the **Mass Entity Config Asset** type, and name it `ME_AudienceConfig`. Then, open it and perform the following steps:

1. In the **Traits** section, add three array elements by clicking the + button three times.

2. From each of the items' dropdown menus, select **Mass Stationary Distance Visualization Trait**, **Assorted Fragments**, and **LODCollector**, respectively.

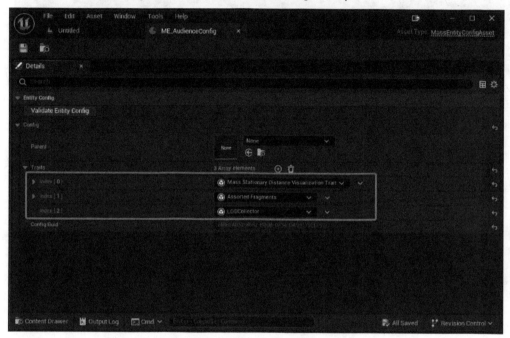

Figure 13.12 – Audience Config data asset

We will now be checking each of them to get some information on how they work.

Configuring Mass Stationary Distance Visualization Trait

This trait is responsible for the visualization of the entity in the world and will need to be properly set in order to correctly show the actor in the level. So, follow these steps:

1. Expand the **Trait** item to show all settings.

2. Set the **High Res Template Actor** property to **BP_AudienceHigh**.

3. Set the **Low Res Template Actor** property to **BP_AudienceLow**.

4. Expand the **Params** section and set the **High** and **Medium** properties to **High Res Spawned Actor** and the **Low** property to **Low Res Spawned Actor**.

5.   Check the **Keep Actor Extra Frame** property.

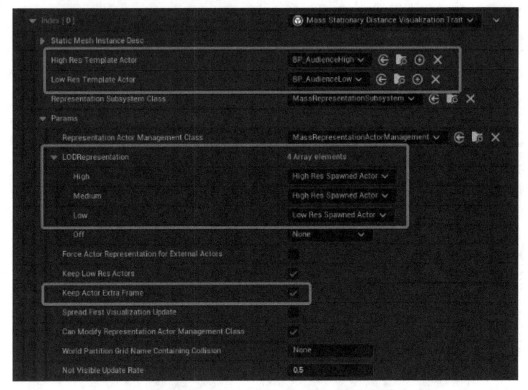

Figure 13.13 – Mass Stationary Distance Visualization settings

As you can see here, we defined a high- and low-resolution template from the previously created Blueprints. Then we used these definitions to set the LOD representation for the entity. Checking the **Keep Actor Extra Frame** flag will help with rendering when transitioning between different LODs.

Configuring Assorted Fragments

The **Assorted Fragments** trait enables you to define an array of fragments that may be required by other traits. To achieve this, open the **Assorted Fragments** trait section and then the **Fragments** section and do the following:

1.   Add two fragments to the array by clicking the + button twice.

2.   Set the first item of the list to **Transform Fragment** and the second one to **Mass Actor Fragment**.

Figure 13.14 – Assorted Fragments settings

Transform Fragment is responsible for storing the entity world transform, while **Mass Actor Fragment** will hold a pointer to the actor that will be used by the visualization trait.

Configuring LODCollector

The entity configuration will need the **LODCollector** trait to facilitate the adjustment between LOD levels. The **LODCollector** processor assesses the appropriate LOD for each entity by considering its proximity to viewers and its relationship to the camera frustum.

It does not require any configuration, so you can leave it as it is.

With the data asset properly configured, we can go on with setting up our system.

Enabling the automatic processor registration

Before creating and testing the gym, we will need to take one more step. At the time of writing this book, the current version of Mass needs some properties to have the **Auto Register with Processing Phases** flag enabled in the Mass settings; this is due to Mass still being in beta status and not yet a final release. Not setting this flag will result in the Mass entities not being visible in the game. This is due to some conflicts with the `MassCrowd` plugin that will be addressed in the future.

To fix this issue, from the main menu, select **Edit** | **Project Settings** and open the **Engine** | **Mass** section. After that, do the following steps:

1.   Expand the **Module Settings** section.

2.   Expand the **Mass Entity** section.

3.   Expand the **Processor CDOs** section.

4.   Search through the array of processors for **MassLODCollectorProcessor**, **MassRepresentationProcessor**, and **MassVisualizationLODProcessor**. Expand the processors and check the **Auto Register with Processing Phases** flag for each of them.

Figure 13.15 – Auto Register with Processor Phases settings

With these settings enabled, we can go on and create a testing gym.

Creating the gym

Since we will be working with a lot of actors, I think it would be fun to take our beloved dummy puppets outside for some fresh air! That's why, instead of using the usual enclosed gym, we will be setting up an open-air environment. Let's start by doing the following steps:

1. Create a new level by using the **Open World** template.

2. Add a **MassSpawner** actor in the level and select it.

3. In the **Settings** panel, set the **Count** property to 50.

4. In the **Entity Types** array, add a new item and expand it.

5. Set the **Entity Config** property to **ME_AudienceConfig**.

6. Add a new item in the **Spawn Data Generators** array and expand it.

7. Set the **Generator Instance** property to **EQS SpawnPoints Generator** and expand its **Query** section.

8. Expand the **EQSRequest** section and set the **Query Template** attribute to **EQS_SpawnEntitiesOnGrid**.

Figure 13.16 – Mass Spawner Details panel

Simulate the level and you should get... something weird!

Figure 13.17 – Wrong position

It seems all the characters are positioned halfway under the floor. This occurs because the **Capsule** component of a character is computed to be centered around the local coordinates of (0.0, 0.0, 0.0). Fixing this issue is trivial and you'll need to adjust a single property of the environment query.

Let's start by duplicating **EQS_SpawnEntitiesOnGrid** and calling it EQS_SpawnEntitiesOnGrid_ ZOffset. Then, do the following steps:

1. Open the newly created asset and select the **SimpleGrid** node.

2. In the **Details** panel, expand the **Projection Data** section and set **Post Projection Vertical Offset** to 120.0.

3. In your level, select the **MassSpawner** actor and change **Query Template** to **EQS_SpawnEntitiesOnGrid_ZOffset**.

If you test the level right now, your characters should be correctly positioned, as shown in *Figure 13.18*:

Figure 13.18 – Right position

Additionally, you should notice that the characters near the camera will be animated, while the ones that are at a distance will be stuck in a cheering position, as shown in *Figure 13.19*:

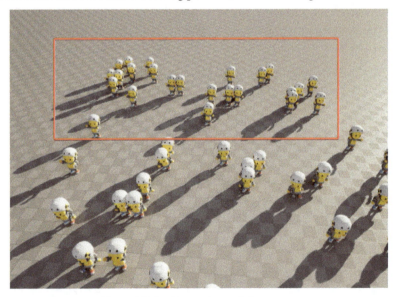

Figure 13.19 – Non-animated characters

This is happening because of the configuration setting we defined in the **Mass Stationary Distance Visualization** trait. I highly encourage you to open the **ME_AudienceConfig** asset back up and tweak the **LODDistance** values – **High**, **Medium**, **Low**, and **Off** – to see how your entities behave.

In this section, I showed you some more advanced traits that compose the Mass framework. We created a new configuration data asset whose task is to manage the LOD of the spawned entities. Then we created a dedicated Mass Spawner that creates high numbers of actors in the level leveraging an environment query.

Summary

In this chapter, we introduced you to the experimental yet highly powerful Mass framework. Starting from the basics, we presented the plugins that make up the whole system. After that, we created a couple of working examples that take advantage of Mass: a simple debug scenario for checking Mass Spawner and a more complex one to get a strong grip on how traits can be combined to handle high numbers of entities.

Using this framework is particularly beneficial in scenarios that require the simulation of crowds, physics interactions, and dynamic entity management. So, if your game needs these types of features, you will highly benefit from using Mass.

In the next – and last – chapter, we will show you another feature that will provide a way to handle and manage various activities and interactions for both AI characters and players. Get ready for an intriguing exploration as things are about to get even more interesting!

14

Implementing Interactable Elements with Smart Objects

In Unreal Engine, **Smart Objects** represent an advanced system implemented to help developers create interactive and context-aware elements within a game. Smart Objects are designed to enhance gameplay by allowing characters – be they player characters or AI agents – to interact with the environment in a more meaningful way through a reservation system. Learning how to use Smart Objects will let you, the developer, create dynamic and interactive environments that enhance gameplay and improve AI behavior for a more immersive player experience.

In this chapter, you will learn the basics of Smart Objects interactivity. We will visit a simple example of how to make AI agents interact with a smart object.

In this chapter, we will be covering the following topics:

- Introducing Smart Objects
- Creating a Smart Object Definition data asset
- Implementing Smart Object logic
- Interacting with Smart Objects

Technical requirements

To follow the topics presented in this chapter, you should have completed all the chapters from *Part 3, Working with Decision Making*, and understood their content. In particular, we will be using part of the code implemented up until *Chapter 11, Understanding the Environment Query System*.

Additionally, if you would prefer to begin with code from the companion repository for this book, you can download the .zip project files provided in the project repository: https://github.com/PacktPublishing/Artificial-Intelligence-in-Unreal-Engine-5

To download the files from the end of the last chapter, click the `Unreal Agility Arena - Chapter 11 - End` link.

Introducing Smart Objects

Smart objects are elements placed within a level that can be interacted with by both AI agents and players. These objects do not contain any execution logic but hold all the necessary information for interactions; additionally, they can be queried at runtime with different methods, such as environment queries.

Smart objects represent a set of activities within a level that can be accessed through a reservation system; if a smart object slot has been claimed by an AI agent, no other agent will be able to use it until it is released.

Presenting the main elements of the Smart Objects framework

Like all Unreal Engine plugins, smart objects are organized into a series of elements, each responsible for a specific task.

The **SmartObject subsystem** is responsible for monitoring all available smart objects within the level and is automatically instantiated in the world when the Smart Objects plugin is enabled. Smart objects are automatically registered with the subsystem for easy access and tracking.

A **Smart Object Definition** is a data asset that holds immutable data shared among multiple runtime instances of smart objects. It contains filtering information such as user-required tags, activity tags, object activation tags, and a default set of behavior definitions used to interact with a smart object. Additionally, a Smart Object Definition features one or more **slots** that can be claimed by players or AI agents to use that specific smart object. Each slot can be positioned relative to its parent, allowing you to define different slots in different positions for the same actor.

To designate an actor as a smart object, you will be using `SmartObjectComponent`; this component will reference a Smart Object Definition asset.

Smart Object Definitions can include one or more **activity tags** that describe the object. They may also feature a **tag query**, which consists of a list of desired tags. This tag query serves as an expression to assess whether the user requesting access to the smart object is permitted to interact with it. As an example, a Smart Object Definition may require a `charging plug` tag that will be usable only by an AI agent that has that exact tag.

Now that you have a quick introduction to the Smart Objects framework, you're all set to dive into your project and start creating your very own smart object.

Starting from the next section, we will be presenting you the basics of using Smart Objects inside Unreal Engine; we won't cover all aspects of the framework as it can be used in many different ways – standalone, or with behavior trees, state trees, and even with Mass. However, by the end of the chapter, you should have a clear understanding of what you can achieve with it.

Time to roll up our sleeves and dive into some code!

Creating a Smart Object Definition data asset

Dr. Markus leaned back in his chair, a satisfied grin spreading across his face as he surveyed the cluttered lab. Tools were scattered everywhere, and the faint hum of machinery filled the air. Across the room, Professor Viktoria was hunched over one of their AI dummy puppets, her brow furrowed in deep concentration.

As the two of them immersed themselves in their work, the idea of upgrading the puppets took shape. The goal was clear: to create puppets capable of not only performing tasks but also fixing themselves when something went awry. This innovation promised to be revolutionary.

With renewed focus, Dr. Markus's fingers danced over the keyboard, typing lines of code that would breathe life into their vision. Each keystroke brought them closer to a breakthrough, and Professor Viktoria's eyes sparkled with excitement as they discussed the potential of their project.

In this section, and in the subsequent ones, you will be creating a gym where AI agents will use a smart object when needed; we will be using the **BP_GunnerDummyCharacter** Blueprint we implemented in *Chapter 9*, *Extending Behavior Trees*, by changing its AI logic. The gunner character will walk randomly, shoot around, and, when the gun jams, try to find a workbench in the level to fix the gun.

To start using the Smart Objects framework, the first thing we need to do is enable the plugins.

Enabling the plugins

To enable the Smart Objects framework, open the **Plugins** window, look for **SmartObjects** and **GameplayBehaviorSmartObjects**, and enable both. You will get a warning as the second plugin is still marked as experimental; go on and restart the Unreal Engine Editor.

Once the plugins are enabled, you will be ready to create your first smart object.

Creating the workbench definition asset

We will now be creating the Smart Object Definition that, as previously mentioned, won't contain any code logic; it will just serve to define the data that will make your soon-to-be-created actor a smart object. To do so, open the `AI` folder and do the following steps:

1. Right-click and select **Artificial Intelligence | SmartObject Definition**.

2. Name the newly created asset `SOD_Workbench` and double-click to open it.

3. In the **Details** panel, locate the **Default Behavior Definitions** array attribute and click the + button to add an element.

4. Set the **Index[0]** element to **Gameplay Behavior Smart Object Behavior Definition**.

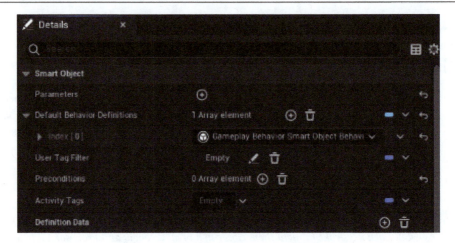

Figure 14.1 – Smart Object behavior definition

As previously stated, a Smart Object Definition contains filtering data that will be used by the system. As we won't be implementing anything complex, we don't need a custom definition, and we will just be using the base one.

It's now time to add a slot, so do the following:

1. In the **Details** panel, locate the **Slots** section and click the + button to add a new slot. Expand the newly created slot.

2. In the **Name** property, insert WorkingPlace.

3. Set the **Shape** property to **Rectangle**.

4. Set the **Size** property to 120.0.

5. Set the **Offset** property to (50.0, 0.0, 0.0).

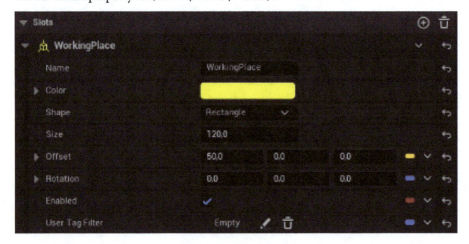

Figure 14.2 – Slot definition

This slot will be the location that can be claimed by the AI agent to fix the gun once it jams and should look like the one in *Figure 14.3*:

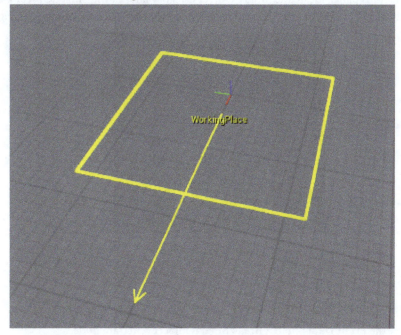

Figure 14.3 – Slot

Keep in mind that you can add as many slots as you deem appropriate. For example, a larger workbench may have more working places or even different ones – for instance, one for fixing guns and one for recharging batteries.

In this section, we have created the Smart Object Definition we will be using to create a Smart Object actor. As you can see, there's no logic involved here because it's handled by the Blueprint we'll be creating in the next section.

Implementing smart object logic

We are now ready to create the Blueprint that will contain the Smart Object definition, making it a fully functional Smart Object. To do so, follow these steps:

1. Open the `Blueprints` folder and create a new Blueprint extending from **Actor**. Call it `BP_Workbench`. Double-click on the newly created asset to open it.

2. In the **Components** panel, add a **StaticMesh** component and select it.

3. In the **Details** panel, set the **Rotation** property to `(0.0, 0.0, -90.0)` and the **Static Mesh** property to **Workbench_Decorated_Workbench_Decorated**.

4. In the **Components** panel, add a **SmartObject** component and select it.

5. In the **Details** panel, locate the **Smart Object** category and set the **Smart Object Definition** property to **SOD_Workbench**. The viewport should now look like the one shown in *Figure 14.4*:

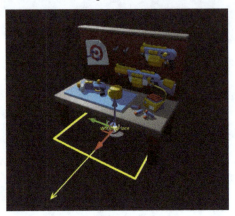

Figure 14.4 – Viewport

Now, open the **Event Graph** panel and do the following steps:

6. Add an **OnSmartObjectEvent** node.

7. Connect the **Event Data** pin to a **Break SmartObjectEventData** node.

8. Connect the **Reason** pin of the **Break SmartObjectEventData** node to a **Switch on ESmartObjectChangeReason** node and click the **Expand** button to show all the switch cases. The graph so far should look like *Figure 14.5*:

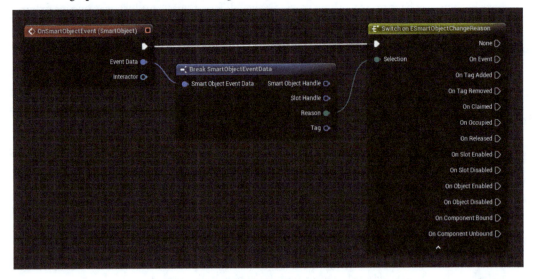

Figure 14.5 – Starting graph

What we are doing here is quite simple; every time we get an event from the smart object, we check what has caused the event. This will serve us to respond to any interaction from the AI agent. Let's continue the code logic by doing the following steps.

9. Connect the **Interactor** pin of the **OnSmartObject Event** node to a **Get AIController** node.

10. Connect **Return Value** of the **Get AIController** node to a **Get Blackboard** node.

Figure 14.6 – Retrieve Blackboard

11. Connect the **Return Value** pin of the **Get Blackboard** node to a **Set Value as Bool** node.

12. Connect the **On Released** outgoing execution pin of the **Switch on ESmartObjectChangeReason** node to the incoming execution pin of the **Set Value as Bool** node.

Click and drag from the **Key Name** pin of the **Set Value as Bool** node. Release and, from the pop-up menu, select **Promote to variable** to create a new variable. In the **Variables** panel, name the newly created variable NeedsReloadKeyName, so that the node in the graph will show a **Needs Reloading Key Name** label. This part of the graph is shown in *Figure 14.7*:

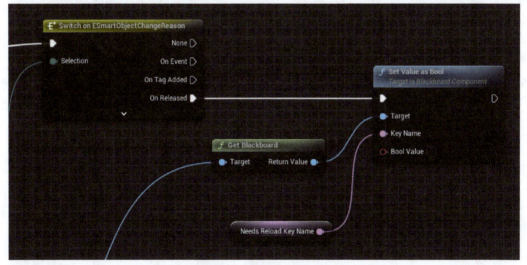

Figure 14.7 – Set Blackboard key

13. Compile the Blueprint and set the variable **Default Value** property to **WeaponJammed**.

In this last part of the graph, we set a Blackboard key value of the AI agent that is interacting with the smart object. We are doing this at the **On Released** event, that is, when the smart object has been claimed, interacted with, and then released to be claimable again. This means we will fix the jammed gun once the AI agent has finished interacting with the smart object.

The Blackboard, and consequently, the **WeaponJammed** key, haven't been implemented yet. In the next section, we will be working on the Blackboard itself and its related behavior tree in order to implement the AI agent.

Interacting with smart objects

In this section, we will finalize the smart object system by creating an AI agent that will make good use of the previously created workbench Blueprint. As stated before, the AI agent will move around and shoot at random; from time to time, the gun will jam, and so the gunner will need to get to the workbench to fix it. The AI behavior will be handled by a behavior tree, and it will be quite straightforward, but it will help us understand how to interact with smart objects.

Let's start by creating a helpful task we will be using in the behavior tree.

Creating the toss coin task

We are now going to create a task that will simulate the gun jamming. This task will be some sort of weighted coin toss and will return a `bool` value – that is, a head or tails result. The weight of the toss will help us in defining how much the gun is prone to jamming. You should already be familiar with behavior tree tasks but, for a quick refresher, you can check *Chapter 8*, *Setting Up a Behavior Tree*.

To implement this task, start by creating a new C++ class extending `BTTaskNode` and call it `BTTask_TossCoin`. Then, open the `BTTask_TossCoin.h` file and add the following block of code:

```
public:
    UBTTask_TossCoin();

    UPROPERTY(EditAnywhere, Category="Blackboard")
    FBlackboardKeySelector BlackboardKey;

    UPROPERTY(EditAnywhere, Category="Task")
    float TrueProbability = 0.5f;

    virtual EBTNodeResult::Type ExecuteTask(UBehaviorTreeComponent&
OwnerComp,
        uint8* NodeMemory) override;
```

The only thing worth mentioning here is the `TrueProbability` property, which will let us weight the result – in our case, the probability of jamming the gun. Now, open the `BTTask_TossCoin.cpp` file and add the following declaration at the top of it:

```
#include "BehaviorTree/BlackboardComponent.h"
```

The constructor is going to be really simple as it will just give a meaningful name to the node. Add the following code to your class implementation:

```
UBTTask_TossCoin::UBTTask_TossCoin()
{
    NodeName = "Toss Coin";
}
```

All the code logic is going to be placed inside the `ExecuteTask()` function. Let's add this block of code:

```
EBTNodeResult::Type UBTTask_
TossCoin::ExecuteTask(UBehaviorTreeComponent& OwnerComp, uint8*
NodeMemory)
{
    const auto BlackboardComp = OwnerComp.GetBlackboardComponent();
    if (BlackboardComp == nullptr)
      { return EBTNodeResult::Failed; }

    const auto RandomNumber = FMath::RandRange(0.0f, 1.0f);
    BlackboardComp->
      SetValueAsBool(BlackboardKey.SelectedKeyName,
        RandomNumber < TrueProbability);

    return EBTNodeResult::Succeeded;
}
```

As you can see, once we have retrieved the Blackboard component, we randomize a `bool` result and set a key value inside the Blackboard itself.

With this class complete, we can now focus on a couple of environment queries that we will need inside the behavior tree.

Creating environment queries

To implement the gunner behavior tree, we will need a couple of environment queries: one for generating a random location for the gunner to reach and one for finding the workbench smart object. Let's start with the first one.

Creating the FindShootLocation environment query

This query will be in charge of generating a set of random locations on the **Nav Mesh** level; basically, we will create a grid of points and later select one of these points as a shooting spot. To do so, open the AI folder and create an environment query by going to **Artificial Intelligence | Environment Query**. Name it EQS_FindShootLocation. Open it and do the following steps:

1. Inside the graph, connect the **ROOT** node to a **Points: Grid** node.

2. Select the newly created node and set the **GridHalfSize** property to 2500.0.

3. Double-check that the **Projection Data | Trace Node** is set to **Navigation**.

Figure 14.8 – FindShootLocation query

As you can see, we are using the nav mesh to look for a location; this will ensure that our agent will be capable of reaching the selected point.

Creating the FindWorkbenchLocation environment query

This second query will be a search for smart objects inside a predefined area. Let's start by creating an environment query (**Artificial Intelligence | Environment Query**) and naming it EQS_FindWorkbench. Open it and do the following steps:

1. Inside the graph, connect the **ROOT** node to a **SmartObjects** node.

2. Add an item in the **Behavior Definition Classes** array property by clicking the + button.

3. Set the item at **Index[0]** to **GameplayBehaviorSmartObjectBehaviorDefinition**.

4. Set the **Query Box Extent** to (5000.0, 5000.0, 500.0).

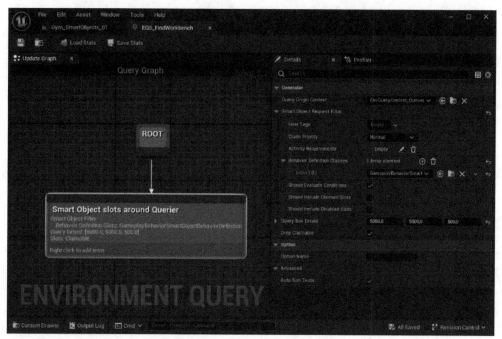

Figure 14.9 – FindWorkbench query

This query will look for any smart object with a behavior definition set to **GameplayBehaviorSmartObjectBehaviorDefinition**; this is the default definition, and it is the one we used inside the **SOD_Workbench** asset. It should be noted that, for the sake of simplicity, we kept things at a very basic level here. I highly encourage you to try implementing a filtering system by using tag filters or by extending your own behavior definition for the workbench or any other smart object you will be adding to the level.

With environment queries finished, we can start implementing the behavior trees, starting from the Blackboard.

Creating the Blackboard

The Blackboard for our AI agent will need to store two locations: one for the shooting target and another for the workbench. Additionally, there should be a flag to indicate whether the gun is jammed. Let's begin by creating a Blackboard and naming it `BB_Tinkerer` to clearly reflect the capabilities of our AI agent. Then, add the following keys:

- A **Vector** named `WorkbenchLocation`
- A **Vector** named `ShootLocation`
- A **Bool** named `WeaponJammed`

With the Blackboard finalized, we are ready to create the behavior tree.

Creating the behavior tree

The behavior tree is going to have two main subbranches that will handle the case when the gun is jammed and when it is not. Let's start by creating a new behavior tree and calling it `BT_Tinkerer`. Open it and set the **Blackboard Asset** property to **BB_Tinkerer**. Then, do the following:

1. Connect the **ROOT** node to a **Selector** node; call it `Root Selector`.
2. Add two **Sequence** nodes to the **Root Selector** node. Call the left one `Shoot Sequence` and the right one `Fix Sequence`.
3. Add a **Blackboard Decorator** to the **Shoot Sequence** node and select it.
4. In the **Details** panel, do the following:

   - Set the **Notify Observer** property to **On Value Change**
   - Set the **Key Query** property to **Is Not Set**
   - Set the **Blackboard Key** to **WeaponJammed**

The graph so far should look like the one shown in *Figure 14.10*:

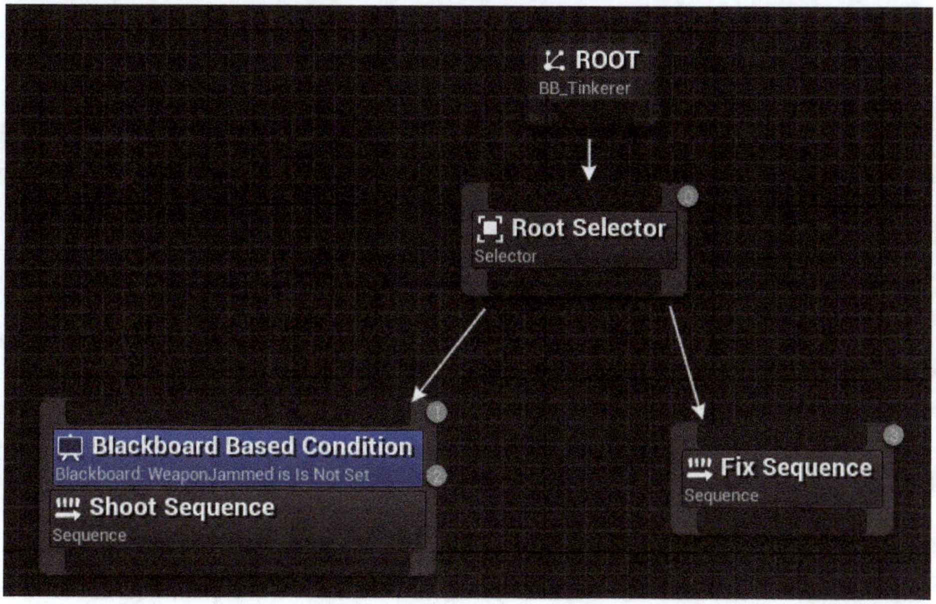

Figure 14.10 – Sequences

Let's now focus on the **Shoot Sequence** section of the graph. Start by doing the following steps:

1. Add a **Move To** task, name it `Move to Shoot Location`, and set the **Blackboard Key** property to **ShootLocation**.

2. Add a **PlayMontage** task to the right of the **Move to Shoot Location** task, name it `Play Shoot Montage`, and set the **Anim Montage** property to **AM_1H_Shoot**.

3. Add a **Wait** task to the right of the **Play Shoot Montage** task. Set the **Wait Time** property to `3.0` and **Random Deviation** to `0.5`.

4. Add a **TossCoin** task to the right of the **Wait** task, name it `Randomize Jam`, set the **Blackboard Key** property to **WeaponJammed**, and set the **True Probability** property to `0.35`.

The **Shoot Sequence** section of the graph should look like the one depicted in *Figure 14.11*:

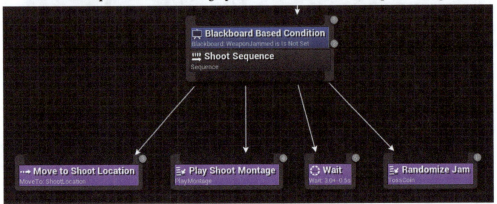

Figure 14.11 – Shoot Sequence

This portion of the graph will move the AI agent to a selected location, start the shooting sequence, wait a short time, and check whether the gun is jammed. However, we need to add a service to make it fully functional. Let's start by selecting **Move to Shoot Location** and adding a **Run EQSQuery** node. Select the query service and, in the **Details** panel, do the following:

5. Name it `Find Shoot Location`.

6. Set the **Query Template** property to **EQS_FindShootLocation**.

7. Set **Run Mode** to **Single Random Item from Best 25%**.

8. Set the **Blackboard Key** property to **ShootLocation**. *Figure 14.12* shows the finalized **Shoot Sequence** section of the graph:

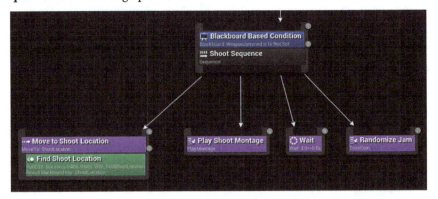

Figure 14.12 – Finalized ShootSequence

This service will execute the environment query that will select one random item from the generated locations, assigning it to the **ShootLocation** property.

With this portion of the behavior tree complete, we can focus on the **Fix Sequence** section of the graph. Follow these steps:

1. Add a **Move To** task, name it `Move to Workbench Location`, and set the **Blackboard Key** property to **WorkbenchLocation**.

2. Add a **PlayMontage** task to the right of the **Move to Workbench Location** task, name it `Play Reload Montage`, and set the **Anim Montage** property to **AM_1H_Reload**.

3. Add a **Wait** task to the right of the **Play Shoot Montage** task. Set the **Wait Time** property to `3.0` and **Random Deviation** to `0.5`.

4. Add a **FindAndUseGameplayBehaviorSmartObject** task to the right of the **Wait** task, name it `Use Workbench`, set the **Query Template** property to **EQS_FindWorkbench**, and set the **RunMode** property to **Single Best Item**. The **Fix Sequence** should look like the one depicted in *Figure 14.13*:

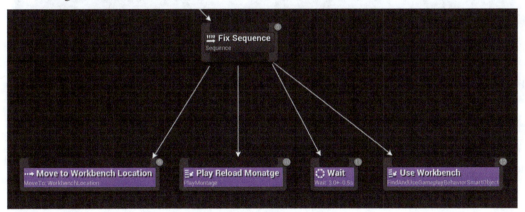

Figure 14.13 – Fix Sequence

You are already familiar with most of the graph, but the last task needs some explanation as it is the most important one. It will find a suitable smart object, claim it, and use it. Then it will release the resource. Although you don't have full control of each of the phases of the smart object as they are executed one after the other, it is quite handy when implementing simple behaviors such as the one we created.

The last thing we need to add is a service that will find the workbench in the level. To implement this, select the **Move to Workbench Location** task and add a **Run EQSQuery** node. Select the query service and, in the **Details** panel, do the following:

1. Name it `Find Workbench Location`.

2. Set the **Query Template** property to **EQS_FindWorkbench**.

3. Set the **Run Mode** attribute to **Single Best Item**.

4. Set the **Blackboard Key** property to **WorkbenchLocation**. *Figure 14.14* shows the finalized **FixSequence** section of the graph:

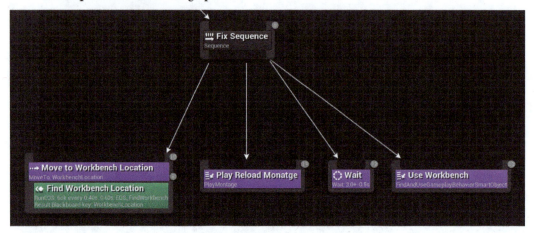

Figure 14.14 – Finalized FixSequence

The behavior tree is now complete; we just need to integrate it into an AI agent and observe its behavior.

Creating the character Blueprints

To finalize our AI agent, we need to create the AI controller and the character Blueprints. Luckily, we have already implemented the necessary classes, and we just need to extend them.

Let's start with the AI controller by doing the following steps:

1. In the Blueprints folder, create a new Blueprint class extending from **BaseDummyAIController**.

2. Name it AITikererDummyController and open it.

3. Set the **Behavior Tree** property to **BT_Tinkerer**.

It's now time to create the character, so do the following steps:

1. In the Blueprints folder, right-click on **BP_GunnerDummyCharacter** and select **Create Child Blueprint Class**.

2. Name the newly created asset BP_TinkererDummyCharacter and open it.

3. In the **Details** panel, set the **AI Controller Class** property to **AITinkererDummyController**.

We are now ready to test things out in a gym.

Testing smart objects in a gym

By now, you should be familiar with the process of creating and testing a gym, so just create a new level of your choice with a **NavMeshBoundsVolume** actor in order to make your AI agent move through the pathfinding system. Then, add a **BP_Tinkerer** instance and a **BP_Workbench** instance to the level and start the simulation.

You should observe the AI agent moving around and shooting. Occasionally, the weapon will jam, prompting the agent to search for a workbench to repair it before returning to its shooting activities.

Figure 14.15 – Finalized fix sequence

In this section, we created an introductory yet fully functional gym that effectively uses smart objects. I highly encourage you to experiment with things a bit by adjusting the jam probability or adding more workbenches and gunners to observe how these changes affect the overall behavior.

Summary

In this final chapter, we learned about Smart Objects, an advanced framework created to assist developers in building interactive and context-aware elements in a game. Smart Objects aim to enrich gameplay by enabling player characters and AI agents to engage with the environment more meaningfully through a reservation system, as we learned in this chapter.

As a developer, you will benefit from using Smart Objects by creating more immersive and interactive gameplay experiences, as environment objects will enable complex behaviors and context-sensitive interactions within the game world. By leveraging this technology, your games will become increasingly captivating and engaging for your players.

Epilogue

In their secret laboratory, Dr. Markus and Professor Viktoria watched as their AI dummy puppets prepared for an epic battle, armed with colorful Nerf guns.

Markus laughed as a bold puppet dashed from behind a lab table, launching foam darts. "We've created the world's first AI Nerf combat league!" He proudly exclaimed.

Viktoria grinned, delighted by their creations' enthusiasm. "Who knew they'd embrace battle so eagerly?"

With the glow of monitors casting light across the lab, Markus suggested, "What's next? A championship tournament?"

"Absolutely!" Viktoria replied, her excitement infectious.

In that hidden haven of innovation, they realized they had built something extraordinary – a whimsical world of creativity and friendly competition, bursting with endless possibilities.

So, this is the end of this book; I hope you enjoyed it as much as I enjoyed writing it!

This is just the start of your exciting journey into the amazing world of artificial intelligence development in Unreal Engine! You might be asking yourself, "*What's next?*". I totally understand that it can feel a bit overwhelming with all the plugins, frameworks, and technologies I presented. To help you out, I've put together a fun new task for you! In the project template, you will find a level called **CaptureTheFlag**. Feel free to dive in and create your very own *Capture the Flag* game featuring the dummy puppets as main characters. Try to combine everything you've learned so far to develop your own AI agents.

Dive into this with all the enthusiasm you can muster and don't be afraid to experiment and play around. Remember, the best experiences often come from letting loose and having a great time, so go ahead and make it your own.

Have a blast!

Appendix – Understanding C++ in Unreal Engine

This appendix has been written to provide you with additional insights, resources, and practical information to enhance your understanding of C++ programming within the Unreal Engine framework. This will help you as a refresher or as a reference throughout the book, just in case you need some help when following the presented code.

This appendix serves as a valuable reference that complements the main content of the book, providing you with the tools and information needed to successfully navigate the exciting world of C++ in Unreal Engine. Whether you're a mid-experienced developer looking to refine your skills or a proficient C++ programmer with little Unreal Engine experience, I hope this appendix will enrich your learning experience!

We will be covering the following topics:

- Introducing basic concepts
- Explaining advanced features
- Exploring core mechanics

Technical requirements

To follow the topics presented in this chapter, you should have some good knowledge about programming, particularly some basic understanding of the C++ language. Additionally, you will need a good understanding of the Unreal Engine.

> **Note**
> This chapter will provide a gentle introduction to the main topics related to using C++ in Unreal Engine. For a more comprehensive exploration of programming in C++ with Unreal Engine, I recommend checking out Zhenyu George Li's book titled *Unreal Engine 5 Game Development with C++ Scripting*, published by Packt.

Introducing basic concepts

If you share my passion for game development and programming, you'll probably agree that writing C++ code in Unreal Engine is both enjoyable and surprisingly accessible. Epic Games has done an excellent job of incorporating features that simplify C++ usage for nearly every programmer.

While it is indeed possible to write standard C++ code in Unreal Engine, leveraging the engine's most widely used features – such as the built-in garbage collector and reflection system – will help you achieve better performance in your games.

In this section, I'll be covering the basic principles behind Unreal Engine C++ features.

Understanding C++ classes

Unsurprisingly enough, an Unreal Engine C++ class is essentially a standard C++ class! If you already have a solid understanding of object-oriented programming in C++, you'll find the environment quite familiar. The process of creating a new C++ class starts by determining the type of object you want to represent, such as an actor or a component. After defining the type, you declare variables and methods in the **header file** (which uses the `.h` extension) and implement the logic in the **source file** (which uses the `.cpp` extension).

The source file works just like any regular C++ file, but the header file allows you to specify additional information for variables and functions that will be accessible to Blueprints inheriting from your class. This also simplifies runtime memory management, as I will explain later.

Let's start by presenting the base types used by the Unreal Engine framework.

The base types

In UE, there are three primary class types that you'll be deriving from during development:

- `UObject`: This is the base class of Unreal Engine, providing core features such as networking support and reflection of properties and methods
- `AActor`: This is a `UObject` type that can be added to a game level, either through the Editor or at runtime
- `UActorComponent`: This is the fundamental class for defining components that can be attached to an actor, or another component belonging to the same actor

Additionally, UE provides the following entities:

- `UStruct`: This is used to create simple data structures
- `UEnum`: This is used to represent enumerations of elements

Unreal Engine prefixes

Unreal Engine class names begin with specific letters; these prefixes are used to indicate the class type. The main prefixes used are as follows:

- `U`: This is used for generic objects that derive from `UObject`, such as components. A good example is the `UStaticMeshComponent` class.
- `A`: This is used for objects that derive from an actor – that is, the `AActor` class – and that can be added to a level.
- `F`: This is used for generic classes and structures, such as the `FColor` structure.
- `T`: This is used for templates, such as `TArray` or `TMap`.
- `I`: This is used for interfaces, such as `IGameplayTaskOwnerInterface`.
- `E`: This is used for enums, such as `EActorBeginPlayState`.

Note that these prefixes are mandatory; if you attempt to name a class that derives from `AActor` without the `A` prefix, you will encounter an error at compile time. This rule applies exclusively to C++ classes; Blueprints can be named without these prefixes. Unreal Engine will hide the C++ prefix once you are in the Editor.

Properties

As you may already know, in programming languages, a **property** refers to a variable that is declared within a class. Adding properties to a C++ class requires some extra attention – specifically, we need to consider whether the code we write should be visible or hidden from the Blueprints that will inherit from our class.

Declaring a property

A property is declared using the standard C++ syntax for variable declaration, preceded by the `UPROPERTY()` macro, which specifies various attributes – such as visibility in Blueprints – and any relevant metadata. As an example, consider the following code:

```
UPROPERTY(VisibleAnywhere, Category="Damage")
float Damage;
```

In the preceding example, setting the `Damage` variable as `VisibleAnywhere` will make the property visible to a Blueprint, but not modifiable. Additionally, it will be logically grouped under the **Damage** category in the **Details** panel.

Property specifiers

As you have already noticed, the UPROPERTY() can include a list of parameters, called **property specifiers**, that will add additional functionality to a property. Some of them are listed here:

- VisibleAnywhere: The property is displayed but cannot be modified in the **Details** panel
- EditAnywhere: The property can be modified in the **Details** panel for both Blueprints and instances placed in the level
- EditDefaultsOnly: The property can be modified in the **Details** panel of the Blueprint but not in instances placed in the level
- EditInstanceOnly: The property can be modified in the **Details** panel of instances placed in the level but not in the Blueprint
- BlueprintReadOnly: The property can be read within the Blueprint but cannot be assigned
- BlueprintReadWrite: The property can be both read and assigned within the Blueprint

Functions

Functions work just like regular C++ functions. Additionally, just like properties, you can decorate them with a macro – in this case, UFUNCTION() – that can contain proper specifiers. As an example, consider the following code:

```
UFUNCTION(BlueprintCallable)
void Heal(float Amount);
```

In this case, the function can be called from a child Blueprint, as it has been decorated with the BlueprintCallable specifier.

For a full list of property and function specifiers, visit the official documentation: https://dev.epicgames.com/documentation/en-us/unreal-engine/metadata-specifiers-in-unreal-engine.

The C++ Header Preview

Unreal Engine comes packed with an incredible inspection tool called the **C++ Header Preview**, which enables you to examine your Blueprint classes as if they were written in C++. To activate this tool, simply navigate to the main menu and select **Tools | C++ Header Preview**, and then choose your desired class. For example, *Figure A.1* displays the header file of **BP_Scrambler** from this book project:

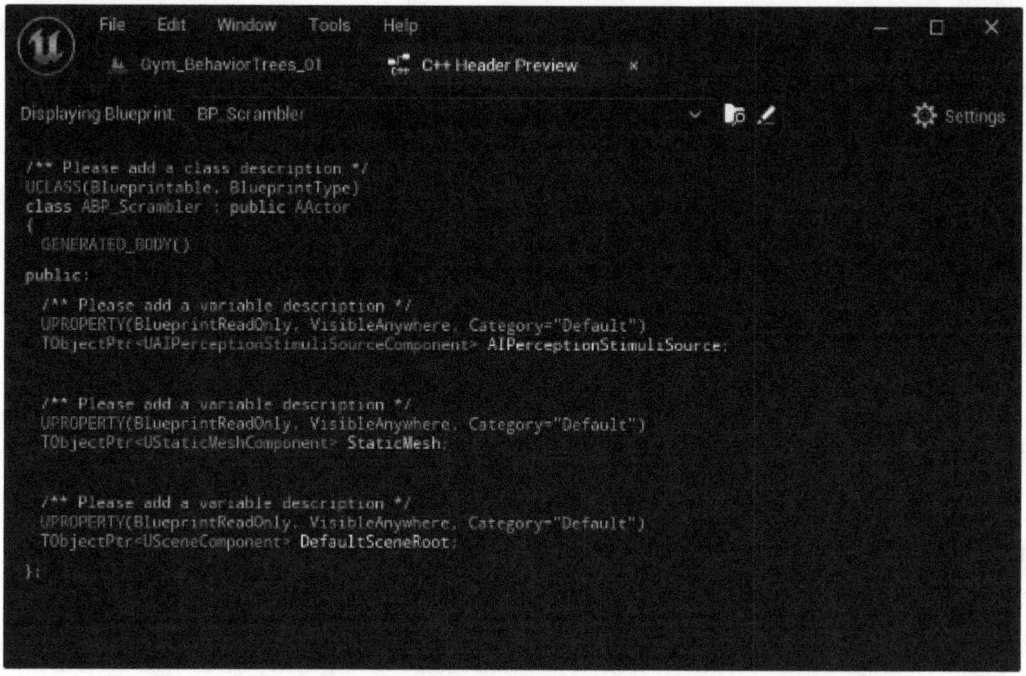

Figure A.1 – The C++ Header Preview tool

This is an amazing tool if already have some Blueprint knowledge and want to be gently introduced to C++ programming.

In this section, I have introduced you to some of the most basic features of C++ in Unreal Engine; in the next section, I will introduce some more advanced features that you may already be familiar with in C++ but that are handled somewhat differently in Unreal Engine.

Explaining advanced features

In this section, we are going to explore how Unreal Engine copes with some of the common features of C++, such as casting and delegates.

Casting

In C++ – and other programming languages, **casting** is the process of converting a variable from one data type to another. It allows you to treat an object as a different type, which can be useful in various situations, such as when working with inheritance or interfacing with APIs. To cast in Unreal Engine, you use the `Cast<T>()` method. As an example, take a look at the following code:

```
APlayerCharacter* PlayerCharacter =
    Cast<APlayerCharacter>(Actor);
```

As you can see, we are trying to cast an `Actor` pointer to an `APlayerCharacter` type.

In Unreal Engine, the `Cast<T>()` function is a safe way to cast pointers to a specific class type, as your code will return `nullptr` instead of crashing when the cast itself fails.

Casting in Unreal Engine should be approached with caution for several reasons. First, it can introduce performance overhead; this can slow down your game if used frequently. Additionally, excessive casting can complicate code readability and understanding. It may obscure the relationships between classes, making it harder for developers to grasp the structure of the code base. One effective way to reduce class dependencies is by using interfaces.

Interfaces

In many programming languages, **interfaces** provide a way to define functions for multiple classes without requiring any specific implementation. For example, your player character might activate different items in different ways. By defining an interface that declares an `Activate()` method, each item implementing the interface will define its own personal logic.

> **Note**
>
> In Unreal Engine, interfaces differ from traditional programming interfaces, as you are not required to provide an implementation for declared functions.

In Unreal Engine, interfaces will need the `UINTERFACE()` macro declaration and two class declarations (in the same file), with two different prefixes. As an example, an `Activatable` interface, with the `Activate()` function, will be written more or less like the following code:

```
UINTERFACE(MinimalAPI)
class UActivatable : public UInterface
{
    GENERATED_BODY()
};
class IActivatable
{
    GENERATED_BODY()

    void Activate();
public:
    UFUNCTION(BlueprintCallable, BlueprintNativeEvent)
    void Activate();
};
```

In this example, the `UActivatable` class declaration is a `UObject` class that contains all the reflection information related to the interface. Being a `UObject` class, it possesses all the features you would normally expect, such as having a name, the ability to be serialized, and support for reflection.

Conversely, the `IActivatable` class declaration is the actual native class used by the compiler to inject virtual functions into your class.

Delegates

In C++, a **delegate** is a type that lets you reference a function, enabling you to call that function indirectly. Delegates are often used for event handling and callback mechanisms, allowing different parts of a program to communicate in a decoupled way. In Unreal Engine, delegates are specifically designed for use with the engine's event system. They allow you to bind functions to events so that when the event occurs, the bound functions are called automatically.

In Unreal Engine, there are several types of delegates:

- **Single-cast**: These delegates only allow one function to be bound to the delegate at a time. They are ideal for scenarios where you want to ensure that only one event handler responds to an event.

- **Multicast**: These delegates can have multiple functions bound to them, allowing multiple event handlers to respond to the same event. They are suitable for scenarios where several components or classes need to listen to the same event and respond accordingly.

- **Dynamic**: These delegates are a type of delegate that can be serialized and are compatible with Unreal Engine's reflection system. They allow you to bind and unbind functions at runtime and can be easily exposed to Blueprints, making them very versatile.

- **Dynamic multicast**: These delegates combine both dynamic and multicast delegates.

If you work in a hybrid Blueprint/C++ project, chances are that you'll be mostly working with dynamic multicast delegates; this will let you expose delegates to Blueprints and bind multiple functions to them. Just keep in mind that, although highly flexible and powerful, using them might impact performance due to runtime binding.

Declaring a delegate

Whenever you declare a delegate, you will use a macro starting with the `DECLARE_` prefix. For example, to declare a dynamic multicast delegate with a single parameter, you will use the following syntax:

```
DECLARE_DYNAMIC_MULTICAST_DELEGATE_OneParam(OnDamageTakenSignature,
float, Amount);
```

As you can see, the delegate declaration defines the delegate name, the parameter type, and the parameter name.

> **Note**
>
> For a full list of the available delegate declarations, check out the official documentation: `https://dev.epicgames.com/documentation/en-us/unreal-engine/delegates-and-lamba-functions-in-unreal-engine`.

Creating a variable of a delegate type

To create a variable out of a delegate type, you will use the following syntax:

```
UPROPERTY(BlueprintAssignable)
OnDamageTakenSignature OnDamageTaken;
```

Note the use of the `BlueprintAssignable` specifier that will make this property accessible to Blueprints.

Subscribing to a delegate

Subscribing to a delegate varies, depending on whether it is multicast or not, as well as whether it is dynamic or non-dynamic. In our example, the delegate is dynamic and multicast, so we will use the following syntax:

```
OnDamageTaken.AddDynamic(this, &ClassName::HandleDamage);
```

The `HandleDamage()` function will be similar to the following code:

```
void HandleDamage(float Amount)
{ /** Function implementation **/ }
```

Calling a delegate

Calling a delegate is quite straightforward, as you will use the `Broadcast()` function for multicast delegates and the `Execute()` function otherwise. Calling a delegate from our example will look like the following code:

```
OnDamageTaken.Broadcast(30.f);
```

Here, `30.f` is the damage taken by the listening object.

In this section, I have presented some of the key features that distinguish Unreal Engine C++ from standard C++ programming. In the next section, we are going to delve deep into some of the most important inner features of Unreal Engine.

Exploring core mechanics

In this section, we will delve into more sophisticated features, such as memory management and reflection, and explain how Unreal Engine copes with them.

Garbage collection

As you may already know, **garbage collection (GC)** is a way to automatically manage memory. In a GC-managed system, once an object is no longer used, it will be automatically removed from memory to free space. This allows you to create a new object and use it, and when you're finished using it, you will simply be good to go. This system is managed by the **garbage collector**, which constantly monitors which objects are still in use. When an object is no longer needed, the garbage collector automatically frees up the associated memory.

While GC is used by many modern programming languages – such as C# and Python – lower-level languages such as C and C++ do not include a garbage collector by default. As a result, programmers must manually track memory usage and free it when it's no longer needed. This process can be error-prone and more challenging for developers to manage. To address this issue, Unreal Engine has implemented its own GC system.

How is GC used by Unreal Engine?

When an object derived from the UObject class is instantiated, it will be registered with Unreal Engine's GC system. This system automatically runs at predefined time intervals – about 30 to 60 seconds – to identify and remove any objects that are no longer in use.

The GC system keeps a set of root objects that are defined to remain alive indefinitely. Additionally, it uses reflection – something that C++ lacks but that Unreal Engine has natively – to examine the properties of an object. This allows the GC system to follow references to other objects and their properties.

If an object is discovered while traversing other objects, and one of those objects is part of the root set, then the object is deemed reachable and remains alive. Once all objects are examined, and if there is no way to reach an object in the root set via references, that object is considered unreachable and marked to be garbage-collected.

When an object is garbage-collected, the memory it occupies is freed and returned to the system; any pointers that referenced this object will be set to null.

> **Note**
> It should be noted that manual memory management – the one you should be used to if you come from a pure C++ background – is still an option in Unreal Engine, but it cannot be used on any UObject derived classes.

Using GC in Unreal Engine

If you have a pointer within a function, you don't need to be concerned about the GC, as pointers inside functions behave like standard C/C++ pointers and do not require any modifications.

Conversely, if you want to have a pointer to an object that you need to persist across frames, you will need to add some small additional code; the pointer needs to be stored as a member variable in your class, and you must add the UPROPERTY() macro before it. That's all you need to do to have the reference that follows be considered by the GC system.

> **Note**
> The UPROPERTY() macro can only be used inside classes that derive from UObject; otherwise, you will have to handle memory manually.

The reflection system

The term **reflection** refers to the ability of a program to inspect its own structure at runtime; this feature is extremely valuable and one of the core technologies used by Unreal Engine, supporting various systems such as the **Detail** panel in the Editor, serialization, GC, and communication between Blueprint and C++.

Since there is no native support for reflection in C++, Epic Games has created its own system to gather, examine, and modify data related to C++ classes, structs, and more within the engine.

> **Note**
> The reflection system also empowers all the Editor's panels, making Unreal Engine's UI highly customizable.

In order to let the system use reflection, you will need to annotate any type or property that you want to expose to the system. This annotation will use macros such as UCLASS(), UFUNCTION(), or UPROPERTY(). Finally, to enable these annotations, you will need to add the #include "AClassName.generated.h" declaration. This declaration is automatically generated when you create a class from the Unreal Engine Editor, so you won't need to worry about it.

As an example, consider the following block of code from the BaseSecurityCam.h file you created in this book project:

```
#pragma once

#include "CoreMinimal.h"
#include "GameFramework/Actor.h"
#include "BaseSecurityCam.generated.h"
```

```
class UAIPerceptionComponent;

UCLASS(Blueprintable)
class UNREALAGILITYARENA_API ABaseSecurityCam :
    public APawn
{
    GENERATED_BODY()

    UPROPERTY(VisibleAnywhere,
              BlueprintReadOnly,
              Category="Security Cam",
              meta=(AllowPrivateAccess="true"))
    UStaticMeshComponent* SupportMeshComponent;

    UPROPERTY(VisibleAnywhere,
              BlueprintReadOnly,
              Category="Security Cam",
              meta=(AllowPrivateAccess="true"))
    UStaticMeshComponent* CamMeshComponent;

public:
    ABaseSecurityCam();
};
```

You will have noticed the presence of the `#include "BaseSecurityCam.generated.h"` declaration and the use of the `UPROPERTY()` macro for the components' declarations.

The following list outlines the fundamental markup elements available within the reflection system:

- `UCLASS()`: Generates reflection data for a class that derives from `UObject`
- `USTRUCT()`: Generates reflection data for a struct
- `GENERATED_BODY()`: Will be replaced with all the necessary boilerplate code for the class type
- `UPROPERTY()`: Informs the engine that the associated member variable will have additional features, such as Blueprint accessibility
- `UFUNCTION()`: Allows, among other things, us to call the decorated function from an extending Blueprint class or override the function from the Blueprint itself

The reflection system is also used by the garbage collector, so you won't need to worry about memory management, as explained in the GC subsection.

Summary

In this appendix, I provided an overview of how C++ is used within Unreal Engine, highlighting its unique features and functionalities. We explored the integration of C++ with the engine's architecture, some of the differences between Unreal Engine C++ and standard C++, and the benefits of using C++ inside the engine itself. We also discussed key concepts such as delegates and memory management, emphasizing their most important peculiarities and features. Additionally, we introduced the C++ Header Preview tool; such an instrument is a must-have if you don't have much experience with C++ for Unreal Engine and you want to make the transition from Blueprints.

Index

A

Other Books You May Enjoy

If you enjoyed this book, you may be interested in these other books by Packt:

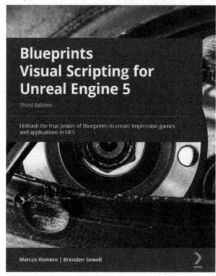

Blueprints Visual Scripting for Unreal Engine 5

Marcos Romero, Brenden Sewell

ISBN: 978-1-80181-158-3

- Understand programming concepts in Blueprints
- Create prototypes and iterate new game mechanics rapidly
- Build user interface elements and interactive menus
- Use advanced Blueprint nodes to manage the complexity of a game
- Explore all the features of the Blueprint editor, such as the Components tab, Viewport, and Event Graph
- Get to grips with OOP concepts and explore the Gameplay Framework
- Work with virtual reality development in UE Blueprint
- Implement procedural generation and create a product configurator

Cinematic Photoreal Environments in Unreal Engine 5

Giovanni Visai

ISBN: 978-1-80324-411-2

- Generate a Master Material to create hundreds of different material instances
- Explore lighting principles and apply them to UE lighting systems
- Evaluate the pros and cons of real-time rendering in the world-building process
- Build massive landscapes with procedural materials, heightmap, landmass, and water
- Populate an environment with realistic assets using Foliage and Megascan
- Master the art of crafting stunning shots with Sequencer
- Enhance visual quality with Post Process Volume and Niagara
- Produce a photorealistic shot using the Movie Render Queue

Packt is searching for authors like you

If you're interested in becoming an author for Packt, please visit `authors.packtpub.com` and apply today. We have worked with thousands of developers and tech professionals, just like you, to help them share their insight with the global tech community. You can make a general application, apply for a specific hot topic that we are recruiting an author for, or submit your own idea.

Share Your Thoughts

Now you've finished *Artificial Intelligence in Unreal Engine 5*, we'd love to hear your thoughts! Scan the QR code below to go straight to the Amazon review page for this book and share your feedback or leave a review on the site that you purchased it from.

https://packt.link/r/1-836-20585-6

Your review is important to us and the tech community and will help us make sure we're delivering excellent quality content.

www.ingramcontent.com/pod-product-compliance
Lightning Source LLC
LaVergne TN
LVHW080112070326
832902LV00015B/2537